To the future of healthcare!

Regards,

Emmanuel Fombu

JLABS SF 2019

THE FUTURE OF HEALTHCARE:

HUMANS AND MACHINES PARTNERING FOR BETTER OUTCOMES

EMMANUEL FOMBU, MD, MBA

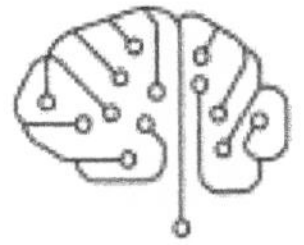

Edited by Dane Cobain
Cover Design by Sean Strong
Photography by Martina Milova
Illustrations by Alex Bashta

CONTENTS

INTRODUCTION

THE GLOBAL HEALTHCARE INDUSTRY is enormous.

In the US alone, the healthcare industry is worth around $1.7 trillion, with nearly 800,000 healthcare companies and 17 million employees[1]. In the UK, the NHS is the world's fifth largest employer[2], and emerging markets are frantically building new healthcare infrastructure to bring them into the 21st century while catering to the boom in the global population.

At the moment, healthcare is too expensive. It's unsustainable. Doctors and patients have the same transactional relationship as a retail customer and a cashier – you show up to your appointment and then pay for the service. The more services rendered, the higher the cost.

I decided to write this book because I strongly believe that our healthcare system in its current state needs to be reformed as it's too expensive and inefficient. With the emergence of new technologies like the internet of things, big data, artificial intelligence and machine learning, healthcare could be substantially improved through personalized drug discovery, prevention, patient education, decreased costs and greater efficiencies. We're at a pivotal point at which people are able – for the first time in history – to share their data and speak openly about their medical histories. And technology is readily available to capture and process this data.

We live in a world where data can help us to make more

---

[1] See: http://bit.ly/2017healthoutlook

[2] See: http://bit.ly/biggestemployer

informed decisions about how to navigate traffic, who to date, what to buy, who to network with and how to better manage our finances. However, when it comes to our personal health and wellness, we have no roadmap. We've made some progress in our understanding of genes, genetic variations and the conditions we're born with, but we don't know enough about what could prolong our health, how illnesses manifest in the body over time and how our daily personal choices can limit or expand our future. We need a map that depicts where we are in terms of our health, with landmarks for risks and opportunities. A path paved by the experiences of those on the road ahead that gives cues about which pathways lead to health and which lead to disease. A GPS that makes it easier to move toward our personal health goals. A new way to look at health and life.

The good news is that people are starting to take their health into their own hands, which is something that's important for every one of us to do. We as people need to take responsibility for our own healthcare. We're all stakeholders in the healthcare system whether we like it or not.

While all of this is happening and the healthcare industry is crawling slowly along, new players are getting into the game with the aim of changing the way we look at it forever. You might have heard of them: Alphabet (Google's parent company), Amazon, Facebook and Microsoft. They're four of the most valuable companies in the world with combined profits of $25 billion in the first quarter of 2017 alone, and they're getting ready to disrupt the healthcare market. After all, they deal in data and view each one of us as a potential customer. Healthcare is the utopian battleground.

The healthcare industry has already started to accept some new technologies because of the efficiencies that they can offer, but in other areas it's woefully behind. The UK's National Health Service is the world's largest purchaser of fax machines[3] despite the fact that faxes are no longer a part of today's society. Meanwhile, many doctors fail to keep accurate notes on their electronic medical record

[3] See: http://bit.ly/faxmachines

(EMR) systems because they'd prefer to spend their time talking to patients.

Even when records are up to date, every hospital has a different EMR system and they don't talk to each other. If a patient starts in one hospital and seeks follow up care at another, their records can be disconnected, which makes the process more difficult and much more inefficient.

Adding insult to injury, the current transaction-based system has misplaced priorities. In America, fee for service hospitals rely on filling their beds with sick patients in order to keep the electricity on and their doors open. How to resolve this issue is up for debate. It's the insurers who are incentivized to improve patient outcomes, because they're the ones who foot the bill. They're saying, "It's too expensive for us. The number of people who are developing preventable chronic illnesses such as diabetes and hypertension is increasing dramatically. They're also living longer due to advancements in medicine and technology. It's not sustainable."

And the insurers have a point. In America, we've been looking at healthcare the wrong way. Healthcare is so expensive because we focus on treatment instead of prevention. In the current fee-for-service business model, there are no major incentives for prevention. On top of that, while people are living longer, they could be sacrificing their quality of life. It's great for people to live longer, healthier lives – but it's not so great when they live longer lives while being bedridden and on various prohibitively expensive therapies for conditions that are largely preventable.

I should note here that I'm not against the appropriate use of medications and procedures. They're often very necessary, but they should be treated as a last resort and targeted to patients who are most likely to benefit from them. Hospitals, doctors, insurers and pharmaceutical companies didn't create this system – they're all victims of the red tape and bureaucracy that leads them astray and takes their eyes off the goal of improving patient outcomes.

On the plus side, insurers are inadvertently pushing us towards a more value-based model by paying hospitals flat fees for each of their customers and penalizing for poor outcomes such as 30-day readmissions. This will also have a knock-on effect for pharmaceuticals. Instead of just selling a pill, pharma companies will have to start thinking beyond the pill. You can see this happening already as more and more pharma companies enter into value-based contracts with payers to cover their medication.

Most articles, publications and practitioners are focused on how tech is changing healthcare. If you see a keynote speaker at a conference, they'll talk about these exciting new technologies and all the amazing artificial intelligence and machine learning algorithms behind them. That focus, in my humble opinion, is slightly misguided – it should be about how tech is delivering value by reducing costs and improving outcomes and quality of life. The subtitle for this book is "humans and machines partnering for better outcomes" – not "humans and machines partnering to do cool stuff in virtual reality."

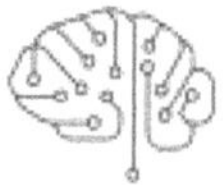

Some people think that new technologies – like the internet of things, big data, artificial intelligence and machine learning – are too complicated, not to mention light years away from any practical use. While the topic maybe complicated, the truth is that we're already using the same technologies every day in other areas of our lives, but the healthcare industry has been as slow as molasses in January to adapt to today's tech world. Whenever you log into Facebook, you're seeing a personalized news feed. When you use Netflix, they're giving you recommendations based on the shows that you've liked in the past, as well as by aggregating data based on what viewers like you are interested in. So why expect anything different when it comes to healthcare?

This isn't science fiction. The future is already here, it just hasn't broken into healthcare. That's partly because there's resistance from people who don't understand AI, machine learning and other technologies. But you don't need to understand it to reap the rewards, just like you don't need to know how Netflix's algorithm works if you're just looking for something to watch. It doesn't matter how it works as long as it gets the job done.

Future healthcare solutions are likely to be similar to existing consumer technology so that it's all just a part of life. Google and Apple are already in the healthcare market. Amazon will soon follow. These big companies are already auditing data like emails, browsing histories and social media posts, so once they fully move into healthcare they'll be well-prepared to crunch the numbers and get a real understanding of what they're looking at. As an advocate for prevention and early intervention in disease management, I believe it will be helpful for the government to put policies in place that will protect consumers and empower them to safely and securely share their data, as it's our responsibility to lead the way towards a better future. The sooner it happens, the more lives will be saved and the better our quality of life will be. It's a moral duty.

Every stakeholder benefits. Insurers will save money. Patients will live longer and have better lives. Pharmaceutical companies will reimagine their business models, thinking beyond the pill and bringing drugs to market more quickly.

The current system is broken. We need to move towards an era of proactive disease prevention and personalized medicine. I vehemently believe that new technologies can be used to educate patients, collect real world data, improve adherence and plan interventions to improve outcomes. We'll no longer need to wait for illnesses to progress to an irreversible point before we start treating them. Imagine a world in which there are fewer 30-year-olds on blood pressure and type 2 diabetes medications because of their poor diets and lack of exercise.

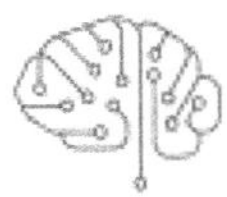

In his 2015 memoir *It's Never Easy These Days*[4], former hospital

[4] See: http://bit.ly/itsnevereasybook

manager Gareth Hollbrooke looks back at his time with the NHS and concludes that, "What may well change is the way in which services are delivered and the roles which healthcare facilities will play in the future. For example, there is likely to be much more visiting of patients at home and attempting to emphasize the preventative elements of healthcare rather than the much more expensive, complicated and, on occasion, the potentially more dangerous practice of placing patients into a hospital setting. An ironic comparison is that evidently in the last part of the twentieth century, the nuclear industry was a safer environment to work in than healthcare."

That can't be possible, right? Surely the nuclear industry can't be safer than the healthcare industry?

Wrong. According to Liam Donaldson, the World Health Organization's envoy for patient safety, "If you were admitted to hospital tomorrow in any country, your chances of being subjected to an error in your care would be something like 1 in 10. Your chances of dying due to an error in healthcare would be 1 in 300."[5] Thanks to a combination of human error, outdated technology and superbugs like MRSA, Donaldson concludes that going to the hospital is almost 35,000 times more dangerous than flying.

Hollbrooke is right, but this change won't only happen within the NHS. The privatized American system, which has been allowed to grow and mutate until patients' wellbeing is no longer at the center of it, will have to change too. It's inevitable.

Just look at how burdensome the current system is. The patient has to make an appointment and then get to the hospital, which can be a challenge when they're not feeling well. If the patient is suffering from an illness that's contagious, they have plenty of time to spread those germs while they wait for hours on end until the doctor is available. Once they're inside the doctor's office, they spend five minutes discussing their symptoms – because that's all the time the doctor has – before heading to the pharmacy to then

---

[5] See: http://bit.ly/riskierthanflying

wait again for their prescription to be filled so it's ready for them to collect. It's inefficient at best and dangerous at worst, putting extra stress on patients at a time when they most need to rest and relax.

That's why there needs to be a change in the way in which these services are delivered. Imagine if your Amazon Alexa could listen to you chat to your physician and create a full transcript. It could store the data on the hospital's EHR system while sending every patient a copy for their records. Let's say that the doctor advises you to diet and exercise and prescribes your medication through Alexa. Amazon could take this order and deliver the drug to you at home on the same day – and then Alexa could remind you to take your medication and encourage you to diet and exercise. Alexa could also teach patients about their disease and explain how their medication works. That helps to increase the likelihood that the patient will comply with the doctor's instructions, which therefore improves their overall outcome. They could even reach their physician on demand using Echo Look from the comfort of their own home.

This is the future of healthcare.

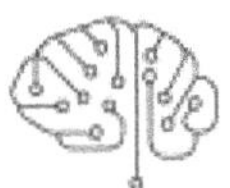

To understand my mission, it might help if I tell you a bit about myself. I'm originally from Cameroon in West Africa. It's bordered by Nigeria to the West, Chad to the northeast, the Central African Republic to the east and Equatorial Guinea, Gabon and the Republic of the Congo to the south. It takes its name from Portuguese sailors who, in 1572, noticed the abundance of ghost shrimp in the Wouri River and named it Rio dos Camarões (Shrimp River) – which eventually became Cameroon in English.

I was raised by my grandmother (Mami Jaco), who was an adamant advocate for education despite the fact that she never spent a day in a classroom. She was a true inspiration, an amazing woman. She sold tomatoes and palm oil at the local market to pay for my mom's tuition.

When I was younger, and a student at a private British boarding school in Cameroon (Sacred Heart College), I was always "that art guy" because I loved literature and reading. Even now, I read a lot and write poetry.

You're probably wondering how I ended up in medical school. I have my mother to thank for that. She wanted the best for her kids, and we grew up knowing she wanted us to go to medical school. Her wish came true and all three of us – myself and my two sisters – moved into the healthcare industry. One of my sisters is a physician in Florida and the other is going through medical school.

I started my medical career in plastic surgery before moving into cardiology after my grandmother suffered from heart failure. But I lost my passion for practicing medicine in a clinical setting once I realized how burdensome it was to work as a physician in a fee-for-service based hospital system. As a result, I embarked on a career journey that's encompassed teaching allied health students, designing clinical trials and carrying out clinical research, founding a health and education focused non-profit organization, advocating for patients with heart disease and working with venture capitalists and start-ups in life sciences, digital health and pharmaceuticals. I also found time to complete an MBA at Cornell University. What can I say? I like to keep busy.

Healthcare is my job, my passion and the thing that gets me up in the morning and keeps me awake at night. Having had the opportunity to be a physician, a caregiver, an entrepreneur and a pharma executive, I took it upon myself to look at healthcare systems both in America and across the world, as well as at some of the ways they can be improved in the digital age.

I was introduced to the concept of big data by Anil Moolchandani, a colleague of mine from Cornell University's Johnson School of Business who serves as the director of engineering programs at SanDisk. Anil's the one who opened my eyes to the true potential of big data, which I define and discuss in chapter five. When I looked at big data and machine learning, I realized that the healthcare industry was already poised for disruption in this new age. But unfortunately, not everyone gets it – they don't realize how much data we gather and what that data allows us to do.

Even now, in the digital age, patients find that their information can't be shared easily between different doctors, especially if they work in different hospitals or different clinics. Instead of being made available to both patients and physicians, the data lives in PDF files attached to emails or physical printouts delivered by fax machines. Even the people who have access through so-called "patient portals" often find that the user experience is poor and the information is limited.

This problem is often called the "interoperability crisis" and it's hurting patients, driving inefficiencies in the system and causing frustration amongst healthcare workers. And you don't have to be a genius to see why – the lack of data-sharing between providers leads to avoidable mistakes and missed diagnoses. Digital technology has the potential to correct all of that, giving patients and the doctors they trust full access to their health information and treatment history.

Marketing departments are using data to deliver more targeted messaging to potential customers. Research and Development departments are using it to develop products that are likely to succeed in our complicated global marketplace. Facebook and Google process huge amounts of data so they can sell advertising space to the highest bidders. But the healthcare industry is falling behind, which could have real consequences for patients. While big internet companies can aggregate data from all over the world, healthcare systems are siloed with no centralized database. This makes it harder to isolate common themes to identify new treatment options, slowing down the healthcare industry as a whole.

There's a better way.

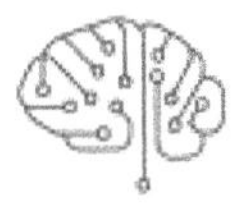

People are already more open with their health information. Medical conditions are no longer a secret, and while mental health is often still stigmatized, the internet has provided a platform for sufferers to connect with each other. People are turning to social networking to be open about their health problems and to use their voices to raise awareness for the conditions that they suffer from.

Take makeup artist and former model Bethany Townsend, who was fitted with colostomy bags after suffering from Crohn's disease since the age of three. We tend to think of young social networkers as filling their feeds with narcissism, food photos and duck-face selfies, but the 23-year-old bucked the trend by sharing a photo of herself as she sunbathed with her colostomy bags.[6]

Bethany's photo was seen by millions of people and received over 190,000 likes and over 10,000 comments on Facebook alone. She credits her husband's support for helping her to beat the stigma of the disease, saying, "When I met Ian, I showed him my bags straight away, but he didn't bat an eyelid. I actually felt like I was overreacting." When interviewed about her viral photo, she added, "I didn't expect this kind of reaction at all. I'm just so glad that it's brought about more awareness of Crohn's disease. If I can inspire or help other people in my position to feel a little more comfortable in their own skin then I'm really happy."

But Bethany's story is far from unique. Nevada mom-of-two Summer VonHesse has been doing something similar, posting photographs of herself that highlight her "flabby, stretched out, loose skin belly."[7] VonHesse hopes to encourage body positivity and self-confidence while highlighting the fact that many mothers lose confidence in their post-baby bodies. Thessy Kouzoukas, the American creative director and co-owner of the fashion brand Sabo Skirt, used Instagram to share two photos that showed the shocking effect that endometriosis had on her body in just a five week period.[8] The condition, in which tissue that behaves like the lining of the womb is found in other parts of the body, affects 10% of women worldwide and sent Kouzoukas into an early menopause.

It's not just patients who are making their voices heard. Just take the case of Joey Daley, whose mother has been diagnosed with Lewy

---

[6] See: http://bit.ly/bethanytownsend

[7] See: http://bit.ly/bikinimom

[8] See: http://bit.ly/endometriosiswoman

Body Dementia. Daley was upset about the lack of real life examples of people dealing with dementia, explaining, "I've searched online for the past three or four years and all I ever find is doctors narrating and the patients are much older. I haven't seen anyone post real footage following someone with dementia and seeing how they handle it and how the caregivers handle it." So with nothing else out there, Daley took the initiative, amassing over 40,000 subscribers and millions of views on his YouTube channel, which follows his mother's progress.[9]

Mental health issues are increasingly talked about, too. Take the case of Madalyn Parker, who emailed her CEO to let him know she was taking time off for mental health reasons. Her CEO, Ben Congleton, replied saying, "I just wanted to personally thank you for sending emails like this. Every time you do, I use it as a reminder of the importance of using sick days for mental health – I can't believe this is not standard practice at all organizations." Parker shared the email trail on Twitter and the story went viral, picking up coverage in mainstream media publications while showcasing the reality of mental health issues and the way they'll be treated in the future.[10]

While I was writing the first draft of this book, two high profile musicians – Chris Cornell and Chester Bennington – committed suicide.[11] Several weeks after Bennington's death, Sinead O'Connor took to Facebook to share a video in which she said she felt suicidal.[12] Interestingly, both Cornell and O'Connor struggled with mental health issues and drug addictions, and they also both released high-profile covers of Prince's *Nothing Compares 2 U*. Prince himself passed away due to an accidental overdose of an opioid called fentanyl.[13]

Tragic as these cases are, they also helped to bring the topic of

---

[9] See: http://bit.ly/joeydaley

[10] See: http://bit.ly/mentalhealthdayemail

[11] See: http://bit.ly/chrischester

[12] See: http://bit.ly/sineadsuicide

[13] See: http://bit.ly/princefentanyl

suicide, self-harm and mental health to the forefront, encouraging productive conversations about how to remove taboos and better treat sufferers.

Mental health issues hit close to home. My editor, fellow author Dane Cobain, suffers from anxiety and depression. While I was writing this book, he sent me plenty of screenshots of friends of his who were open and honest about their mental health issues, including Heba El-Husseini, who used to work for an online counselling platform and who shared the following update: "Girls, be careful with Norethisterone. I have never experienced such rapid onset of depressive symptoms and honestly it sent me loopy for the best part of a week before I had to discontinue and just have a period from hell regardless. Obviously this won't happen to everyone and probably works fine for a lot of people, but messing with your hormones can be a dangerous game so just be careful everybody." When asked for her opinion on the future of healthcare, El-Husseini said: "It's super cool that mental health is being prioritized by innovators. I think medication should be a last resort rather than an immediate band aid – because ultimately that's all it is for many people. It's not a sustainable fix for a long-term problem."

The world that we live in is becoming increasingly diverse and inclusive. It's okay for people to express themselves, and it's also okay for them to talk about their medical conditions. In fact, these conversations generate huge amounts of data that could help medical practitioners and researchers to develop new treatments and medications. For scientists to develop drugs and devices that provide positive outcomes for the patient, it's essential for platforms to exist that allow patients to share their thoughts on what matters most to them. The patients and caregivers will provide structured and unstructured data that can be processed using AI and machine learning to develop products that address the needs of patients – and not just the needs of scientists.

And yet when I talk to other people in the healthcare industry, I find myself fighting to convince them that this world is possible, even though it's already here. Millennials and internet natives, the next generation of doctors and patients, are already used to talking about their health, and they're entering the workplace in huge numbers. By 2030, hyper-connected, tech savvy millennials will make up 75% of the workforce.[14] The stage is set for a more open approach to healthcare.

I want to make people open up to themselves and say, "It's okay to talk about my health conditions." Because it is.

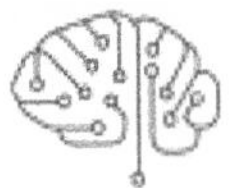

I don't like to drive, but I live in New York City where there's a lot of traffic and I commute to New Jersey for work. With all of the time spent in the car commuting, I've developed a newfound love for audiobooks. Part of my inspiration for this book came while I was stuck in traffic and listening to *The Digital Doctor* by Robert Wachter. He was talking about electronic healthcare systems and the daily struggles doctors face thanks to the use of digitalized medical records, an approach that's currently standard across the industry. Doctors are notoriously slow at adopting new technologies, so much so that it's become a cliché. This struck a chord with me as I could recall the frustrations I encountered earlier in my career which pushed me away from clinical medicine. I was spending more time entering information into a computer and trying to ensure drugs were covered by insurance plans than actually talking to my patients.

Wachter is a member of a new breed of healthcare theorists who see the true potential of technology, and his writing is descriptive and often humorous, packed full of anecdotes and a joy to read. I tried to take inspiration from that because I don't want my book to be another dry medical tome that reads like a college reference book. I want it to be an enjoyable read that entertains, educates and gives you a glimpse of the future that patients and healthcare professionals have to look forward to.

---

[14] See: http://bit.ly/millennialworkforce

There are plenty of people out there who are doing a great job of ushering in the future of healthcare. One of those is my friend Dr. Zubin Damania, who's better known as internet celebrity ZDogg MD, the comedy rapper with over 18 million views on YouTube.[15] Then there's San Diego's Eric Topol, a cardiologist, geneticist and digital medicine researcher who works for the hospital system. He's released a bunch of books and consulted for companies like WebMD and Medscape, where he was appointed Editor in Chief.

These thought leaders are doing fantastic work when it comes to bringing the future of healthcare to doctors and practitioners, but I want to do something a little different. I've worked all over the healthcare industry, from practicing as a clinician to working in pharmaceuticals and dealing with insurance companies. I'm a member of a small group of healthcare specialists who can understand every stakeholder involved – and who knows that no one talks to each other.

Every one of us is invested in healthcare, and at some point in our lives every one of us is a patient. We all have a vested interest in ushering in the future of healthcare, whether we work in the industry or whether we simply want the best possible treatment for ourselves and for our loved ones.

When you write a book, people start to ask you who your audience is. The truth is that we're all stakeholders in healthcare, from the day we're born until the day we die. This book isn't just for healthcare professionals. This book is for everyone – you included.

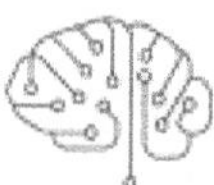

We live in a world of big data and analytics. A world in which everything is connected. A world that moves so quickly that, by the time you've finished this sentence, there will have been around

---

[15] See: http://bit.ly/zdoggyoutube

220,000 new Facebook posts, 23,000 new tweets, 7,000 new app downloads and about $9,000 of sales through Amazon.[16] That's a lot of data.

Technology and the data that it allows us to collect is set to change the face of healthcare forever. But technology can't do it alone, which is why the new generation of stakeholders will need to move with the times and accept the possibilities that technology can bring to life.

The rise of technology and the shift towards the future of healthcare is set to be accompanied by a change in the way that we look at mental and physical health conditions. Like Crohn's sufferer Bethany Townsend, we'll no longer allow health conditions to stigmatize us – and that means we'll be free to collect data on them to identify the best way to prevent or treat them.

This shift is already underway, but it needs two key things if it's to continue and make the future of healthcare a reality. First, it needs leaders – people within the healthcare industry who can spread information and work together towards a common goal. And second, it needs support from people like you, the reader.

In the end, it's down to all of us to make this happen. There needs to be a groundswell from the public that's accompanied by a new generation of tech-savvy practitioners who use new tools to improve medicine for the masses.

The public needs to demand change. If you make your voice heard – by supporting the future of healthcare, lobbying your congressman and demanding better healthcare through new technologies – the government will listen.

The power is in your hands. Only you can usher in the future of healthcare.

Emmanuel Fombu

Emmanuel Fombu, MD, MBA
April 20th, 2018

---

[16] See: http://bit.ly/realtimenetstats

CHAPTER ONE: THE FUTURE OF HEALTHCARE

"THE ABILITY TO CULTIVATE MASS AMOUNTS OF DATA TO ZERO IN ON CURES AND NEW TECHNIQUES TO ENHANCE THE SPEED OF HEALING [IS WHAT EXCITES ME MOST ABOUT OUR DIGITAL FUTURE]."

– BILL PEACOCK, COO OF CLEVELAND CLINIC

LET ME TAKE YOU on a journey through space and time.

We're heading into the future. It's not as close as the day after tomorrow and it's not as far as 4,000 AD when humanity has colonized space and moved way beyond the solar system. It's somewhere in between, a future that's within our reach thanks to the exciting technology that's already around today.

We're going to follow the stories of three people to see the future of healthcare in action, and in the next chapter we're going to take a look at how the future of healthcare contrasts with where we are today.

MEET DIANA

Diana is a 53-year-old library assistant with two loving children and a husband who works in real estate. The family doesn't really need the money, but she loves her job and the smell of books is like a drug to her. The thought of all of those books – all of that *knowledge* – gets her out of bed in the morning.

Well, her virtual assistant helps, too.

At precisely 7:15 AM every weekday, she's woken by the chirpy tones of Amazon's Alexa. Alexa knows that 7:15 AM is the best time for Diana to wake up based on her body's circadian rhythm and the metrics it receives from her smartwatch. When she gets up earlier – or later – she typically gets less exercise, which can then affect her blood pressure.

Diana's story began on a typical morning. Alexa woke Diana by slowly raising the lights and playing classical music, which helped to bring her gently to consciousness instead of waking her with a jolt. She sighed and stretched, noticing as she did so that her shoulder was stiff and there was a dull ache in her foot.

"Good morning, Diana," Alexa said. "You slept well. I make it just under eight hours, which is within the healthy range for a woman of your age. Don't forget to take your medication."

"Thanks, Alexa," Diana replied. She pulled herself out of bed, walked over to the dressing table where her atenolol was already waiting and swallowed it down with a glass of water. Then she walked over towards her wardrobe.

"Diana," Alexa said. "It looks like you're limping. Is something wrong?"

"Ah," Diana replied. "It's nothing. I must have stubbed my toe."

"Are you sure?"

"No," Diana admitted. So Alexa asked her a checklist of questions, routing the answers through a set of algorithms to determine a preliminary diagnosis.

"Diana," Alexa said, "I think you ought to go to the hospital. It might be nothing, but I'm worried about potential infections and I lack the equipment to fully scan you. Would you like me to call a taxi?"

"Sure thing," Diana said, and before she knew it she was in the back of an Uber and worming her way through the streets towards the hospital. She felt a little silly about bothering the doctors with a sore foot, but she remembered the last time she'd ignored Alexa's advice, when she'd thought she had a cold and ended up being diagnosed with mild pneumonia.

At the hospital, the waiting room was almost empty. Diana checked in at the desk and was quickly ushered in to speak to the doctor.

"Hi, Diana," he said. "My name's Dr. Barlow. TREWS tells me that we'll need to run some further tests, I hope that's okay."

"TREWS?"

The doctor nodded. "Our targeted real-time early warning system," he said. "It uses machine learning to analyze data from patients to pick up on symptoms that we might otherwise miss. Your creatinine level is a little high."

"What does that mean?"

"Let's find out," the doctor said.

High creatinine levels can be caused by multiple different conditions, including sepsis and chronic kidney disease, so Dr. Barlow and his team carried out a battery of tests. Based on Diana's medical history and the analysis of other patients with similar attributes, Barlow and his team knew that kidney disease was an unlikely suspect, and his virtual assistant calculated that there was a 93% chance that they were dealing with sepsis.

Barlow took a chance and started Diana on antibiotics, then told her she could go back home as long as she wore a hospital health tracker and sent in a digital blood sample using one of their disposable kits.

Two days later, Diana's creatinine levels were back to normal and she'd never felt better.

"Good morning, Diana," Alexa said. "You're looking well today."

MEET OMAR

Omar had always wanted to be a physician. His father was a doctor, his grandfather was a surgeon, and his great grandfather had been a medic during the Second World War. It was a family tradition.

One morning, Omar was woken up as normal by a gentle vibration from his health tracker. He wore it on his wrist so that it could track his vitals and help him to maintain a steady sleep cycle. But on that morning, it detected something unusual, so it routed the data through the cloud and crunched the numbers, comparing Omar's vitals to those of other people of his age and ethnicity.

Omar's phone went off, so he answered the incoming Skype call. As soon as the video feed loaded, he could see who it was – Dr. Groves was calling, which usually wasn't good news.

"Hi Omar," Dr. Groves said. "How are you?"

"I'm okay, doc," Omar replied. "What's up?"

"Just a quick one," the doctor said. "Your tracker tells me that your blood pressure is a little high. It's not necessarily something to worry about, but I think we'd both feel a lot better if we checked it out. I was wondering if you could stop by today to give us a blood sample?"

"Sure thing. Let me just get dressed."

"Great," Dr. Groves said. "I'll see you later."

Omar hopped on the back of his bicycle and rode to the local medical center where Dr. Groves worked. The doctor was busy on his daily rounds, most of which were carried out remotely, and so Omar was greeted by a nurse called Joy who directed him into a side room and showed him how to use the bloodwork machine.

"It's easy," she said. "Just put your thumb in the hole and you'll feel a quick prick from the needle. It'll be over before you know it."

"What happens then?" Omar asked.

"The machine will destroy the needle and prepare a new one for the next patient," Joy replied. "Meanwhile, we'll analyze the blood and get Dr. Groves to take a look at it. If you'd like to take a seat in the waiting area, he'll come and get you shortly."

Omar went to sit down in the waiting area, figuring he'd play on his phone to pass the time. He was on his third level of Angry Birds when Dr. Groves called him through into his office.

"Good to see you again," the doctor said. "Take a seat. This shouldn't take long."

Omar did as he was told. "What's the news?" he asked.

"Nothing serious," the doctor said. "At least, not at this stage. But I would like to keep an eye on your blood pressure. If we tackle it now, we can stop you from experiencing problems in later life."

"What do I need to do?"

"The usual," the doctor replied. "Keep an eye on your weight, eat a healthy diet and get some exercise. I'd also like you to refrain from smoking and to limit your alcohol intake."

"I don't drink."

"Yes you do," the doctor said. "The data never lies. But don't worry, I'm bound to keep your information confidential. It's just to help us to make a diagnosis."

"Okay," Omar said. "Is there anything else I need to do?"

"Yes," the doctor said. "I'm going to transfer a prescription to your smartphone. Please take this to the pharmacist and tell them I sent you."

"Thanks!" Omar said. He shook the doctor's hand and said goodbye, then jogged along the street to the pharmacy. There, he showed them his phone and they scanned the code that the doctor had sent him, and then they invited him out back to collect his prescription.

Omar had never taken medication before – because the doctors were reluctant to hand it out unless there was no other alternative – and so he was keen to see the process. His parents said it had all been different in their day, and he had no idea what to expect.

But the process was quick and painless. The pharmacist explained it to him while she was running it through the machine.

"It's easy," she said. "When your physician ran your bloodwork, it ran through Medbot, our machine learning software, to generate a genetic profile. We can pick out common strands between different patients and tailor the medication to get the dosage right. Medbot has crunched the numbers and now we're running your prescription through the medication printer."

"The medication printer?"

"Yep," she said. "It's like a 3D printer that prints medication. We're running off a course of pills that are tailored to your genetic and environmental needs. Cool, huh?"

"Yeah," Omar said. "I guess so. How long will I have to take them?"

"That's something you need to discuss with your physician," the pharmacist said. "But we usually try to get people off them where possible. We find that around 90% of people are off their medication again within a year."

"So I guess I'd better get some exercise," Omar said.

The pharmacist smiled at him. "I guess you're right," she said, sweetly. "Ah, look. Your medication is ready."

MEET BUDD

Budd was a war veteran. He'd served his country as a younger man, picked up an injury in the war and suffered from PTSD in later life. At 82-years-old, he was feeling his age.

Despite his infirmity, Budd still lived at home. He was still healthy enough to make it up and down the stairs, and while he'd started to withdraw from the local community, he still made it out of the house a couple of times a week to play bridge or to scout for bargains in the shops.

But Budd's family worried about him, so they decided to invest in the latest Alexa Health kit from Amazon. The internet giant had reimagined its business model and offered same day delivery and installation on all of their health products, as well as a monthly consultation with one of their licensed practitioners. The kit included microphones and motion sensors, which were hooked up to the cloud and stored in a centralized database. His kids could keep an eye on it, and even Budd had to admit that it was good to have some backup. He said it reminded him of the old army days when they'd all worked as a unit.

For the family, it provided peace of mind because if their father changed his routine then they'd receive a notification. They were worried about him and his lack of a social life. Since his wife had died, he'd grown increasingly insular, and those rare trips to town were his only chance to socialize, especially because they didn't visit him as often as they used to do. If something did happen, Budd could call for help and the family would automatically receive an alert. Meanwhile, his doctor could track his water intake and whether he'd remembered his medication. Not that there was much chance of him forgetting – not with Alexa reminding him.

The first warning signs came a couple of months after the system was first installed. Alexa noticed that Budd had stopped his visits to town and that he was no longer doing his daily exercises. In fact, he was barely moving at all, spending most of the day in front of the television.

"What's wrong, Budd?" Alexa asked. "You haven't been yourself."

"I have no energy," Budd replied. "I'm tired all the time, and besides – what's the point?"

"I don't like the sound of this, Budd," Alexa said. "Better call the doctor."

Budd agreed and Alexa put the call through, and the doctor asked a series of questions from the PHQ-9 test before arriving at a tentative diagnosis.

"The PHQ-9 test is designed to screen for both the presence and the severity of depression," the doctor said. "Based on the answers you gave and the activity we have from your Alexa Health device, it seems likely that you're suffering from moderate depression."

"Depression?" Budd said. He thought about it for a moment. "No, not me. You must be wrong. I'm not that type of person."

"There *is* no type of person," the doctor replied. "Just trust me on this. Let's start to treat it and see what happens."

"You mean medication?"

"Yes," the doctor said. "To begin with at least, although it's not a long-term solution. I'll process the prescription as soon as our call is over. I'd also like you to work with Alexa. She'll remind you to take your medication and ask you some routine questions each day just to make sure that everything's as it should be. She can also book your follow-up appointments. I'd like to speak to you again in a couple of weeks to see whether there's been any improvement, okay?"

"Okay," Budd said. He waved as the doctor cut the call.

Budd's medication arrived within the hour. As promised, the doctor ordered the medication electronically and Amazon Pharmacy fulfilled the order by printing it at the local warehouse and shipping it same-day with one of their delivery drones. He was able to start taking it straight away, although Alexa requested permission to monitor his vital signs in case of an unexpected reaction.

Budd was more than happy to oblige, and he also asked Alexa for regular reminders to take his medication and to get a little exercise.

"I'm sure I can help with that," Alexa said. "We're also going to pair you up with another patient on LIVYANA, our peer support platform. Somebody your age with the same symptoms."

"How does that work?"

"It uses an algorithm," Alexa replied. "Based on your interests and lifestyle. Ah, here we are. Her name is Jessica and she's 79-years-old. Would you like to meet her?"

Budd said he would, and over the coming weeks he became close friends with his fellow sufferer. In the meantime, Alexa played his favorite music and TV shows, changing up the tempo depending upon his mood, and she even controlled the curtains to let more light in. She encouraged Budd to work out a little more and to start going back into town again, and he agreed to it. After all, he wanted to impress Jessica, and he'd grown to look forward to their calls.

By the time he went back to his doctor, he was feeling much better, although the doctor warned that they'd need to keep him under surveillance for a few months to make sure his condition kept improving.

By focusing on heading off the condition before Budd's depression got worse, his physician was able to avoid expensive counselling and long-term prescriptions, saving hundreds of dollars in consultation and medication. He also found someone to take his mind off his wife – and his loneliness.

THE FUTURE MEETS THE PRESENT

Each of these stories has its basis in real-life modern medicine, and the technology already exists to make bespoke healthcare experiences possible. The laws of supply and demand are enough to ensure that the medical industry is already moving towards the widespread adoption of technology to improve its levels of service.

According to one report from BI Intelligence[17], digital disruption has created an $8.7 trillion opportunity in the healthcare market. New technologies face some resistance – including regulation, staff buy-in and privacy concerns – but the positives outweigh the negatives and digital healthcare is slowly breaking down barriers as people see the value that new technologies have to offer.

Let's take a look at what would happen to the same patients within the current system.

---

17 See: http://bit.ly/disruptioninhealthcare

MEET DIANA (AGAIN)

Diana's story is loosely based on a Tedx talk by Suchi Saria.[18] Saria is a professor of computer science and health policy, as well as the director of the Machine Learning and Health Lab at Baltimore's Johns Hopkins University. If there's anyone who's qualified to talk about machine learning and its impact on healthcare, it's Saria.

Diana (not her real name) is a real patient of Saria's acquaintance, a 52-year-old woman who came to the emergency room to complain about a footsore. When doctors inspected her, they saw no major cause for concern but wanted to monitor her in case the footsore was infected. After a couple of days, she started to develop symptoms consistent with mild pneumonia, but the usual antibiotics didn't help and her condition worsened. By day six she'd developed tachycardia and by day seven, she was struggling to breathe.

Diana went into septic shock and her body was in crisis. It took

[18] See: http://bit.ly/suchisaria

all this for doctors to become concerned and to transfer her into intensive care, but much of the damage had already been done. Her kidneys started to fail. Then her lungs started to fail. Finally, three weeks after visiting the emergency room, Diana died.

After a short investigation, hospital workers identified sepsis as the culprit. Sepsis can trigger a negative inflammatory response and lead to multiple organ failure, and it's the 10th most common cause of death in the US.[19] It's easy to treat but difficult to identify, and the doctors at the hospital didn't recognize Diana's symptoms until it was already too late.

Diana's story also demonstrates some of the exciting potential uses of new technology. TREWS, which used machine learning to analyze patient data, is similar to technology developed by Stanford University and iRhythm Technologies. Their 34-layer convolutional neural network (CNN) can detect arrhythmias in ECG signals better than a cardiologist[20], and the technology could be coupled with low-cost ECG devices to enable more widespread use of the technology as a diagnostic tool – especially in places where access to a cardiologist is limited or otherwise difficult.

---

[19] See: http://bit.ly/sepsisstats

[20] See: http://bit.ly/ecgproject

MEET OMAR (AGAIN)

Omar's story is fictional, but much of the technology that's featured is either on the market or in development. For example, genome sequencing (the process of identifying the precise order of nucleotides that makes up an organism's DNA) is already big business and advancements in the last twenty-five years have reduced the costs of sequencing substantially.[21] In Omar's story, the results of the genome sequencing on his blood test were used to customize the 3D-printed pills that were prescribed as a short-term fix while he addressed his behavior to stop heart disease from being a problem in the future.

We don't have that technology yet, but we're already using genome sequencing to provide a readout of all six billion letters in a person's DNA. One study found that 22% of people had genetic variants in genes that are associated with rare, inheritable diseases[22],

[21] See: http://bit.ly/dnasequencingcosts

[22] See: http://bit.ly/dnatestmit

and two volunteers had genetic variants known to cause heart rhythm abnormalities despite cardiology tests making no discoveries. This means that they could develop heart problems in the future, although it's not a certainty. Either way, it's something that both the patients and their doctors can be on the lookout for.

In our current healthcare system, Omar would be placed on medication and expected to keep taking it for the rest of his life. High blood pressure is a serious problem that affects 30% of women and 32% of men.[23] It's the main risk factor for strokes and a major risk factor for heart attacks, heart failure and kidney disease. Yet despite this, more of a focus is placed on giving out medication than on stopping high blood pressure from becoming a problem in the first place.

On top of that, approximately 62,000 unnecessary deaths occur every year in the UK alone thanks to poor blood pressure control – something that could be avoided with the intelligent use of new technologies and personalized healthcare. Meanwhile, with global obesity rates doubling since 1980[24], high blood pressure is a constant worry for patients and physicians alike.

Omar's story highlights a challenge for the current healthcare system. At the moment, with our transaction-based model, we focus mainly on short-term fixes like medication. In the future, patients like Omar will be identified before their health issues become a serious problem, and through a mixture of personalized healthcare and wearable devices to monitor their behavior, they'll be able to adjust their lifestyle to avoid long-term problems.

Disease prevention makes sense. After all, it's like servicing a car. Regular maintenance and check-ups with a mechanic will help the car to stay on the road for years to come and retain its resale value, simultaneously saving the owner thousands of dollars in unnecessary repairs. On the other hand, if you fail to take care of the car, it'll cost an arm and a leg to repair it when the engine is

---

[23] See: http://bit.ly/bloodpressurestats

[24] See: http://bit.ly/obesityfactsheet

damaged due to neglect. Putting clean oil in a bad engine won't fix the problem because it's already too late.

Healthcare is the same, and prevention is better than cure. Technology is giving all stakeholders the opportunity to maintain the engines (i.e. our bodies) so that we don't suffer from diseases later on in life that could have been prevented or managed early.

MEET BUDD (AGAIN)

Parts of Budd's story are based on *Aging in Place and the Internet of Things* by Claire Mitchell and Vaughn Shinall from Temboo.[25] Their company allows programmers to create code for any internet of things application, so they're in the perfect place to examine how the technology is revolutionizing contemporary healthcare.

Unfortunately for Budd, the current healthcare system is nowhere near this advanced when it comes to catering to people with mental health issues. Medication is often looked at as the first

---

[25] See: http://bit.ly/aginginplaceandiot

and only treatment option, despite the fact that the side effects can often be unpredictable. Counselling sessions work well, but they also take up a huge amount of time and resources, which is why in the future, we'll be able to save costs by talking to each other in virtual support groups.

Depression is relatively common but underdiagnosed in older adults, especially when they suffer from chronic illness and require home healthcare or extended stays in hospital. Loneliness can also be a problem, especially if they're living alone, with estimates of depression amongst older adults ranging from 1-5% of the population.[26]

Meanwhile, technological advances, new research and the delivery of modern healthcare standards have reduced mortality from diseases and extended life expectancies in both developed and developing countries.[27] We're living longer lives, but length and quality are two very different metrics. People who might have died from their condition can now survive, but they face the emotional costs of long-term treatment and medical surveillance. For example, a patient who had a liver transplant will need to continue immunosuppression treatment and may face further complications in the future. The emotional costs of these conditions are often overlooked, which isn't in keeping with the holistic healthcare we'll see in the future.

Still, Budd's story does a lot to show us how elderly care will function, as well as how new technologies will help us to treat complex mental illnesses such as anxiety and depression. These illnesses are typically brought on by a wide range of factors, which can make them difficult to diagnose and to treat. Human beings are naturally inclined to think in absolutes, but machines have no such problem. This will come in useful when treating mental health problems and disorders such as autism which work on a sliding scale with differing levels of severity.

---

[26] See: http://bit.ly/cdcdepression

[27] See: http://bit.ly/emotionaldimensions

HEALTH SCORING

We all know that health insurance costs play a huge role in the quality of care that an individual can expect to receive. The better the insurance, the better their treatment. But I foresee a future in which all of our health information – from lab work, wearable devices, the internet of things, etc. – will be used to generate a health score. This score will change in real time and you'll be able to improve your score by sharing more data. It's similar to the credit scoring system that banks use to decide whether to lend someone money.

My hypothetical health scoring system will determine how likely you are to develop different diseases and will affect how much you pay for health insurance. It works like modern day car insurance and interest on bank loans. The more you exercise and live a healthy lifestyle (thus reducing your risks of disease), the better your health score will be and the cheaper your premiums will be for both healthcare and life insurance.

That brings me to a true story. It starts with a friend of mine we'll call Niba. Niba is 35-years-old with a good job. He has a wife and three kids. He also has a history of high cholesterol, high blood sugar, diabetes, binge eating disorder and anxiety. His anxiety stems from the fact that his father died from a heart attack at the age of 39 and Niba doesn't want to suffer the same fate.

Niba loves his wife and kids and wants to live a long, healthy life so he can continue to support them. I know that because he told me – like I said, we're friends. He knows that he has to eat healthily and get some exercise, but he believes he's too lazy and undisciplined. Instead, he takes prescription pills for weight loss, anxiety, cholesterol and hypertension, as well as a number of different vitamins. I saw him take his pills once and he literally took a dozen of them – and that was only in the morning. As he took them twice a day, that equated to 24 pills every day. That's 168 a week and 8,760 every year.

Last time I saw him, Niba was also on a "diet". We went out for lunch together and he ordered a salad. Afterwards, he picked up fried chicken and mashed potatoes from a takeout place "for his family". Again, he's my friend. I knew he was taking it home for himself because his family was away on holiday, as he'd reminded me earlier that day. He does the same thing on business trips – he'll buy chocolates as gifts for his family and they'll never make it home. And even when we asked him to join us to play some sports, he still refused to exercise.

Meanwhile, Niba visits his doctor every couple of months to calm his anxiety by having every test he can think of. Nuclear stress tests, calcium scores, molecular diagnostic tests, etc. You name it, he has it. The pills, unnecessary tests and visits to the doctor all come at a huge cost to the healthcare system.

Imagine if Niba continues at this rate for the next 35 years. Imagine how much that will cost the healthcare system. And Niba has the power to avoid it by dieting and taking some exercise instead of just depending on medication.

Unfortunately, even though his doctors told him to diet, exercise and take some pills, Niba ignored the rest and only took the pills. Pills are the easy option. But pills are often a quick fix, and outside of genetics, the majority of his health problems are caused by his lifestyle choices.

With a health scoring system, Niba would be able to see an objective overall ranking of his health. He'd pay more for health and life insurance because his score is low, but if he made those lifestyle changes then his score would improve and he'd pay less for insurance while simultaneously improving his health and the health of his family. Technology could be brought in to coach him and to encourage him along the way. He could even confide in Alexa or his peer support group buddies so that he didn't have to hide his eating habits and his lack of exercise.

WHAT'S NEXT?

Now that you've seen the future of healthcare, it's time to take you back to the present. This process is a journey – after all, you can't usher in the future overnight – and for a journey to be successful, you need a map.

This chapter showed you where the destination is. The next chapter will show you where we are now, and the rest of the book will share the directions we'll need if we want to get there.

It's time to take a look at the current system. Buckle up.

CHAPTER TWO: THE CURRENT SYSTEM

"OUR HEALTHCARE SYSTEM SQUANDERS MONEY BECAUSE IT IS DESIGNED TO REACT TO EMERGENCIES. HOMELESS SHELTERS, HOSPITAL EMERGENCY ROOMS, JAILS, PRISONS – THESE ARE EXPENSIVE AND INEFFECTIVE WAYS TO INTERVENE AND THERE ARE PEOPLE WHO CLEARLY PROFIT FROM THIS CYCLE OF CONTINUED SUFFERING."

– PETE EARLEY, JOURNALIST AND AUTHOR

THE MODERN HEALTHCARE SYSTEM is very different to the one that we'll see in the future. It's a complicated, confusing system that puts profits and transactions ahead of actual healthcare benefits, and there are all sorts of misconceptions around the industry. It's so inaccessible that many people try not to think about it until something is wrong, and they often delay seeking lifesaving treatment because they'd rather wait it out and hope for the problem to fix itself. Even when they decide to see a professional, it takes an average of 3.4 weeks to secure a primary care appointment.

A recent study from the US government in the journal Health Affairs found that about 5% of the population – those who are most frail or ill – accounts for nearly half of healthcare spending in any given year. Meanwhile, half of the population has little-to-no healthcare costs, accounting for 3% of spending.[28]

Of the total of $3.35 trillion spent in 2016, hospital care accounts for the largest share at about 32%. Doctors and other clinicians accounted for nearly 20%, while prescription drugs bought through pharmacies accounted for around 10%. Meanwhile, the out-of-pocket cost paid directly by consumers increased as the number of people covered by high-deductible plans continued to grow.

So why exactly are American hospitals so darn expensive? An answer has been provided by an unlikely source: US investigative comedy show *Adam Ruins Everything*. When faced with the claim that "the hospital is expensive but it's worth it" if you get the best treatment, host Adam Conover says, "No it isn't."[29]

Why exactly is that? Well, according to Conover, "A big part of the reason is that American hospitals overcharge patients massively. [A] neck brace is worth $20 but the hospital [charges] $154." This time, Conover explains, the problem isn't politicians but rather the Chargemaster, "a secret document full of insane prices that hospitals use to charge us whatever they want."

What follows is a trip through the history of medical billing that I've reproduced below – although I'd recommend watching the video for the full impact:

A hundred years ago, hospital pricing was pretty simple. [They took] the cost of providing care and [added] a little on top to make a profit. One amputation cost five bucks, so [they charged] you $6.50. But after the rise of insurance companies, hospital billing got complicated, in part because gigantic corporations demanded gigantic discounts. So to please these powerful insurance companies, hospitals cooked up a plan. [They made] up a really high, fake price and then gave them a discount off that. And in less than a century, healthcare prices went from reasonable to nonsensical.

These crazily inflated prices are kept in the hospital's Chargemaster. [Some people say], "I only pay my premium. It they wanna rip off my insurance companies with their fake prices, what do I care?" If you ever lose

---

[28] See: http://bit.ly/healthaffairsstudy

[29] See: http://bit.ly/adamruinseverything

insurance you'll care, because here's the really evil part. If you don't have insurance, you actually get charged these fake prices.

Even if you're insured, you can get billed Chargemaster prices if you go out-of-network. And anything can be out-of-network. The hospital you go to, the equipment used to treat you, even the doctors you see. Hospitals make a ton of money overcharging out-of-network patients. It's a real cash cow and we all get milked.

Every hospital has its own Chargemaster. A treatment that costs $7,000 at one hospital could cost a hundred grand down the road, and you can't comparison shop when you're dying. Plus, since your insurance company faces inflated costs, that can trickle down to you in the form of higher premiums.

We need to go to the hospital, so they have no incentive to change how they do business, and politicians have spent decades arguing over how to pay the bill instead of asking why the bill is so high. Until they do, we're stuck with this system.

Then there are people like Martin Shkreli, "the most hated man in America". The businessman and investor attracted widespread criticism when his company obtained the manufacturing license for an antiparisitic drug called Daraprim and raised its price from $13.50 to $750 per pill – despite the fact that it's available as generic medication elsewhere in the world for $0.10 per dose.[30]

Unfortunately, there will always be people like Shkreli who are too arrogant for their own good and who put individual profit ahead of overall gain. Thanks to Shkreli, the negative image that consumers already have of healthcare has been cemented. He's the poster child for people who believe that the costs of healthcare are out of hand. The problem is that his actions have distracted everyone from the real reasons why healthcare costs are so high. Some might argue that it costs a lot of money to carry out research and development, and it takes an average of thirteen years to bring a new drug to market.

That said, it's certainly true that the healthcare industry is full of wastage. A great analogy comes to us from Toronto, where an

---

[30] See: http://bit.ly/shkrelimosthated

Etobicoke senior built a set of stairs into his local community park.[31] Adi Astl explained that he contacted his councilor to ask for an access point so that the elderly could safely access the garden. They provided an estimate of $65,000 - $150,000, so Astl took matters into his own hands and built a wooden set of stairs for $550. It didn't take long until he was contacted by the council and asked to remove it – as well as being told that he could be charged under a city bylaw.

This highlights one of the problems with the healthcare industry. Stakeholders are working against each other, instead of working together – and there's too much reliance on an inflexible system, instead of on personalized healthcare for every patient. This drives costs up and leads to the medical equivalent of physicians building wooden stairs and then pulling them down again. Meanwhile, the patient still has no way of getting into their metaphorical garden – but they've paid through the nose to go full-circle.

In the fee-for-service system, the idea is that the more services you provide, the bigger the bill. It rewards doing more, instead of rewarding conservatism and taking a more targeted approach. Of course, if a patient actually needs more services to improve their health, conservatism should take a back seat. The argument here is against doing more just for the sake of doing more with no measurable value in return. Author Charles Hugh Smith explains: "In effect, fee-for-service is open-ended. It's like going to an auto mechanic and agreeing to pay for whatever services he deems necessary, at whatever price he chooses, with no penalties to the provider if the service is poor." Walter Cronkite cut to the heart of the matter by saying, "America's healthcare system is neither healthy, caring, nor a system."

Like the night in Game of Thrones, the healthcare system is dark and full of terrors. Winter is coming unless we usher in the future of healthcare. Luckily, you don't have to be Jon Snow to understand what's happening. Let's start by figuring out who has a vested interest in healthcare. It's time for you to meet the stakeholders.

---

[31] See: http://bit.ly/tomrileystairs

MEET THE STAKEHOLDERS

Stakeholders are people who have a vested interest in something. For a company, that could include everyone from staff and suppliers to local institutions and the customers who buy their products.

When it comes to healthcare, there are more stakeholders than you might imagine. In fact, you could argue that everyone in the world is a stakeholder in the medical system. Here are just a few of the key stakeholders in the modern healthcare industry, along with a brief description of where they stand in relation to the others. Later on in this book, we'll look at each of those stakeholders in turn and examine how the future of healthcare will affect them.

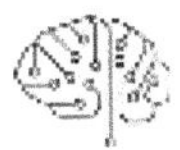

PATIENTS: We're all patients at some point in our lives, which means that we're all stakeholders in the future of healthcare.

PHARMACEUTICAL COMPANIES: These are the companies that are responsible for developing and distributing drugs and medication.

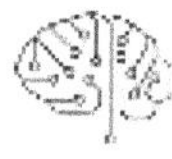

PHYSICIANS: The doctors who carry out the treatments. Physicians' roles vary drastically – some specialize and some are generalists.

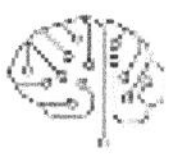

HOSPITALS: From British GP surgeries to American medical institutes and field hospitals in the middle of warzones, if it provides treatment to patients then it's a hospital.

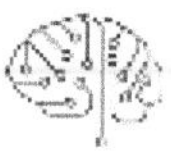

INSURERS: Somebody needs to pay the bills, and that job often falls to insurers.

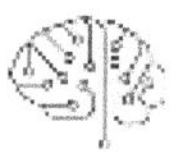

THE GOVERNMENT: It's the government's job to take care of its people and it does that by passing laws and regulations which affect the healthcare industry and subsidizing costs where possible.

HEALTHCARE IN THE US

The current system is unbelievably inefficient. We spend more than $3 trillion every year, and for what? Americans have poorer short-term and long-term outcomes than other developed countries that spend much, much less. It's the system itself which is broken, which is why we need to usher in a new future if we want to fix it.

The way that we diagnose and treat patients will have to change, and so will the way that patients and their doctors interact with each other.

Here's what US healthcare looks like at the moment.

A BROKEN SYSTEM

In 2015, the Commonwealth Fund published its *U.S. Health Care from a Global Perspective* report[32], which drew upon data from the Organization for Economic Cooperation and Development to compare healthcare spending, supply, usage, costs and outcomes across 13 high-income countries. Those countries were Australia, Canada, Denmark, France, Germany, Japan, the Netherlands, New Zealand, Norway, Sweden, Switzerland, the United Kingdom and the United States, in case you're curious.

You probably won't be surprised by their findings:

THE US SPENDS THE MOST ON HEALTHCARE: The US spent 17.1% of its GDP on healthcare in 2013. That's almost 50% more than second-placed France and almost double the UK. Spending per person was equivalent to $9,086.

PRIVATE HEALTHCARE COSTS ARE HIGHEST IN THE US: Americans rank second behind the Swiss on costs for out-of-pocket healthcare and ahead of the rest of the world in all other categories. US insurance costs average at $3,442 per person, more than five times runners-up Canada.

PUBLIC SPENDING ON HEALTHCARE IS HIGH BUT COVERS FEWER RESIDENTS: The US was the only country covered to not have a universal health care system. Public spending on healthcare averaged at $4,197 per person, behind only Norway.

---

32 See: http://bit.ly/ushealthcareglobal

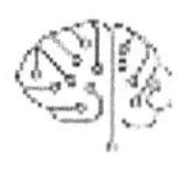

HIGHER SPEND BUT FEWER VISITS: America has a below average number of physicians and inhabitants are less likely to visit doctors or hospitals. This is despite the fact that they pay more overall for the privilege.

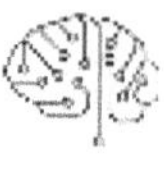

AMERICANS ARE BIG CONSUMERS OF MEDICAL TECH: Unsurprisingly, Japan also ranked high. But America came on top which shows that the will is there to adopt new technologies.

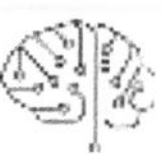

AMERICANS USE MORE DRUGS: America ties with New Zealand to score the highest number of prescription drugs (2.2) per person.

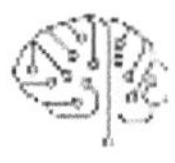

HEALTHCARE COSTS ARE HIGHER: This might sound obvious given the rest of these facts, but this statistic covers all aspects of healthcare. For example, MRI scans, CT scans and heart bypass operations are all much more expensive in the US. So is medication – in 2010, all countries in the report had lower prices than the United States.

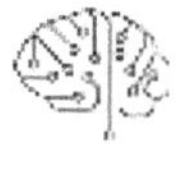

SOCIAL SERVICES SPENDING IS LOW: The US spends the least on social services, such as retirement and disability benefits and supportive housing. This echoes the fact that a holistic approach to patient support is needed, instead of simply medicating and then discharging them.

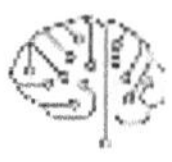

OVERALL HEALTH IS LOW: The United States might spend the most money, but the population as a whole falls below the international average. Americans have the lowest life expectancy at birth[33], the highest infant

---

[33] See: http://bit.ly/healthspending

mortality rate and an above average number of chronic diseases.

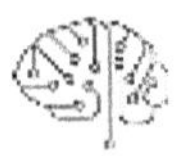

CANCER CARE IS ABOVE AVERAGE – BUT HEART DISEASE AND DIABETES CARE IS OVERLOOKED: America is winning the war against cancer, with US mortality rates lower – and declining faster – than those in other industrialized countries. But the opposite is true of heart disease and diabetes care, two illnesses which are largely preventable if caught early enough.

Despite healthcare spending in the United States exceeding that of other high income countries, it sees poorer results on many key health metrics including life expectancy, quality of life and the prevalence of chronic conditions. The US was the only country in the report without a publically financed universal healthcare system, and it still spent more public dollars on healthcare than all but two other countries.

The system is broken – and policymakers aren't doing much to fix it.

THE THREE TYPES OF POLICY

When it comes to coverage, there are three different types of policy in the United States:

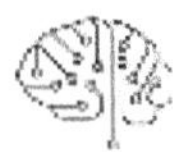

FEDERAL POLICY: Overall policies that span the country, such as Medicare and legislation from the Center for Disease Control.

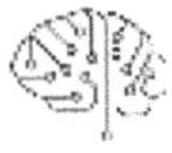

STATE POLICY: Secondary policies that are unique to specific states, such as Medicaid (which tends to vary from state to state).

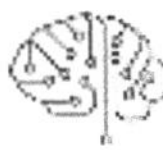

LOCAL POLICY: Tertiary policies at county level that cover local services such as mental health clinics.

THE HISTORY OF HEALTH INSURANCE

Health insurance isn't new. It isn't even old, it's ancient. Back in ancient China, physicians were often paid only if their patient got better. In England, Henry I introduced a series of sweeping healthcare reforms and at least one physician, John of Essex, was being paid on a daily retainer. Mark Twain is on record as saying that throughout his childhood, his parents paid the local doctor a flat fee of $25 per year.

Until the end of the First World War, American medical costs were surprisingly low, and it was the cost of the lost labor that did the damage. One study from 1919 found that lost wages from illness were four times larger than the medical costs, which is why people started to take out "sickness" insurance.[34] Health insurance itself was still relatively rare.

But that was before new advances in medical technology made better healthcare standards available – but only to those who could afford them. This was accompanied by rising incomes and a surge of rural families moving to urban areas. When the depression kicked in, prepaid hospital plans began to grow in popularity, helping to keep hospitals open during the leaner years while giving patients peace of mind.

As the depression wore off, health insurance plans became more popular amongst employers, most notably with Blue Cross/Blue Shield. They realized that keeping employees healthy was paramount for the smooth running of the company. These insurance policies were supported by the government thanks to tax breaks for companies and employees who signed up for insurance. By 1968, 75% of Americans had some form of private coverage.

In 1973, the Health Maintenance Organization (HMO) Act was

---

[34] See: http://bit.ly/1919study

passed, and commentators predicted that one of two things would happen. Either the entire healthcare industry would come crashing down or a revolution would take place in which care and coverage would be provided to every citizen. Neither happened.

Instead, the healthcare system continued to fragment, moving away from a network of collaborative relationships towards the competitive marketplace we see today. Unfortunately, this competitiveness has reduced the potential for data sharing and productive partnerships and pushed costs up to such an extent that between 2000 and 2005, the cost of family coverage increased by 73% while wages increased by only 15%.

The result is the imperfect system we see today, and it seems as though no one is actually happy with it. A 2004 Harris poll confirmed this when it asked people to rate how well 15 different industries served their customers. Managed care companies tied with tobacco companies, insurers and pharmaceutical companies ranked just above oil companies and while hospitals ranked towards the middle, they still lost out to banks, computer companies and airlines.

So when we talk about the history of health insurance, it's basically downhill all the way from the ancient Chinese and John of Essex.

HOW HEALTH INSURANCE WORKS

Taking out health insurance is a little like placing a bet that one day, you're going to get ill. Nevertheless, it's a necessity in the existing system.

The interesting thing about health insurance is that people are actually rewarded if they fail to claim on it. If you're treated by a healthcare facility and you pay them without claiming on your insurance, it'll keep your premiums down.

Thanks to increasing diversification and innovation – both in the healthcare industry and in the world as a whole – there are multiple different types of health insurance plans for people to choose from.

The two most popular plans in the United States are Health Maintenance Organization (HMO) and Preferred Provider Organization (PMO) plans.

HMOs are the more basic plans, with lower premiums but more restraints. They give subscribers access to certain doctors and hospitals within their network, but they won't cover them if they go outside of that network to see a specialist. HMOs also typically cover a limited number of visits, tests and treatments.

PPOs have higher premiums and typically include a minimum deduction, but they're also more comprehensive than HMOs. While they encourage you to see a doctor or hospital within their network of providers, you can go outside of the network and still receive some coverage.

Insurers make more money when people stay out of the hospital. They have a vested interest in keeping their members well to the best of their ability because not doing so cuts into their profits.

INCENTIVIZING ILLNESS

Under the fee-for-service model, hospitals, physicians and pharmacies are incentivized to see the patient as often as possible. The more often they see them, the more money they'll receive, which creates an unhealthy culture in which we overmedicate and fail to address the root cause of health problems until it's already too late.

The United States has approximately one practicing doctor for every 400 people. That means we don't have enough doctors, and it drives up the cost of healthcare due to the laws of supply and demand.

Doctors expect – and deserve – to be well-compensated for the work that they put in. After all, the average doctor graduates medical school with $183,000 of debt, and they've also lost out on ten years of earning power.[35] While I won't deny that the vast majority of doctors get into the field because they want to help their fellow man, they also need people to get ill and to pay for their services. If everyone was healthy, they'd be out of a job and saddled with too much debt to ever pay off.

CUTTING COSTS

Medication is expensive. It's not unusual for psychiatric medication to cost hundreds of dollars per month, and that's before we start to look at specialized treatment such as hyperbaric oxygen therapy for rehab patients.

And with insurers footing the bills for these treatments, they're understandably keen to cut costs. That's why many of them hire a pharmacy benefit manager, who's tasked with the job of keeping an eye on the company's pharmaceutical costs. When it comes to injectables and infusions, they can cost a thousand dollars for every dose, and it's the pharmacy benefit manager's job to find cheaper alternatives. They're usually paid a salary with a commission based on how much money they save, which incentivizes them to cut costs wherever possible – sometimes to the detriment of the patient's health.

Insurance companies often go one step further, using algorithms and human insight to categorize different medications based on how much they cost. They reason that patients get what they pay for – so if you're on a lower plan, you'll be entitled to fewer medications than someone else who's paying a higher premium.

This, of course, creates a problem. In the fee-for-service model, it's not about giving the best possible treatment for the patient. Instead, it's about managing the healthcare process as a series of transactions. Physicians don't decide how best to treat a patient – the

---

[35] See: http://bit.ly/hardtoheal

insurance company does.

Worse still, many medical insurance providers use complicated language and hard-to-decipher eligibility clauses. Patients get confused very quickly, and it's easy to understand why. Sometimes the professionals get confused, too.

It's often not even the insurer who's making the decisions. Many of them outsource their processes to companies like IBM, which means that, technically speaking, IBM is the care management company.

Within the current system, there's a real need for patients to more closely vet their medical insurance – and to do it *before* they get ill. Shopping for insurance is like buying a car. You can buy a basic model off the shop floor, but you can also customize it to include extra safety features and better mileage. It might cost you more up front, but you get what you pay for. I don't know many people who wouldn't be willing to pay a little more if they could afford it, if it meant a better standard of healthcare for themselves and their loved ones. And therein lies the problem. Not everyone can afford to.

WHY IS US HEALTHCARE SO EXPENSIVE?

The United States has the highest healthcare costs in the world, but it doesn't have outcomes to match it.[36] In fact, the US ranks low compared to other developed countries on all sorts of key health metrics, including life expectancy, prevalence of chronic conditions and the mortality rate from heart disease – which is the leading cause of death in the United States.[37] So why is US healthcare so expensive? And where does the money go?

According to a report from The Commonwealth Fund, which draws upon data from the Organization for Economic Cooperation

---

[36] See: http://bit.ly/highspending

[37] See: http://bit.ly/leadingcauses

and Development, "Healthcare spending per person is highest in the US not because Americans go to doctors and hospitals more often but because of greater use of medical technology and healthcare prices that are higher than other nations."[38] In fact, Americans pay more for healthcare services and products, as well as for pharmaceuticals.

People are quick to point the finger at healthcare CEOs, and perhaps there's a reason for that. According to Axios[39], the CEOs of 70 of the largest US healthcare companies have earned $9.8 billion in the seven years since the Affordable Care Act was passed, and their earnings have grown faster than that of most Americans. They also note that pay packages don't give influential healthcare executives incentives for controlling spending – and that the 70 companies in the study bring in $2 trillion of annual revenue.

But there are plenty of other reasons for high healthcare costs, including higher administrative costs and an increased use of expensive technologies, as well as a lack of transparency throughout the system. Americans might be visiting their doctors less than citizens of other countries, but they pay a lot more for the privilege. And insurers can't cover everything, which brings me on to my next subject.

MANAGING THE PATIENT POPULATION

This is a controversial topic, but it's also a topic that needs to be covered. Talking about keeping the costs down is all well and good, but it's not as simple as skipping brand name medication. There are three main factors that impact the way that insurers and healthcare practitioners are able to manage the patient population.

---

[38] See: http://bit.ly/ushealthcareglobal

[39] See: http://bit.ly/healthceopay

COMPLEX CARE

Complex care is an umbrella term that's used to refer to people who require long-term or continuous care. These patients are often suffering from chronic illnesses and disabilities such as brain damage, epilepsy, multiple sclerosis, muscular dystrophy or learning disabilities. Because of their more complex needs, these patients often require specialist care.

Economist Vilfredo Pareto famously created the Pareto principle – or the 80/20 rule – to describe a system in which there's an unequal relationship between inputs and outputs.[40] For example, 20% of doctors recruit 80% of patients in clinical trials. That's how I first discovered it, and I did some work to understand why.

In the healthcare industry, 20% of the patients take up 80% of the resources. The United States has roughly 320 million residents and a $1.4 trillion healthcare budget. That 20% – 64 million Americans – takes up just over $1.1 trillion of that budget. The 20% uses $17,500 each, while the 80% averages out at $1,090. Those numbers aren't entirely accurate, of course, but they do help to illustrate how the Pareto principle can be applied to healthcare.

Organ transplants are another big ticket item that cost insurers and patients a lot of money. Some hospitals specialize in nothing but transplants. Long-term care is also expensive, especially if the patient is dealing with a serious and chronic condition. The longer they live, the more expensive their healthcare becomes – and in our current system, the majority of our healthcare costs kick in during the last few years of our lives.

MEDICAL MALPRACTICE

Doctors are on the defense. These days, many physicians and

---

[40] See: http://bit.ly/theparetoprinciple

facilities ask patients to sign documents to protect them from litigation, and you may have seen an episode of *House* or *Grey's Anatomy* where a surgeon refused to operate because it was too risky. It can be safer to watch someone die than to take a 40% risk on a kill-or-cure course of action.

Honestly, doctors sometimes make mistakes. It's what makes them human, and it's one of the reasons why artificial intelligence, big data and new technologies will change the way we practice healthcare. Human beings make mistakes every day, and doctors aren't immune – they just try to minimize the damage and to do the best they possibly can. But by augmenting their human abilities with artificial intelligence – humans and machines partnering for better outcomes – we can cut down on human error and reduce the overall cost of healthcare by giving malpractice attorneys fewer opportunities to file a lawsuit.

This brings us to one of the key reasons why humans and machines need to work together – instead of against each other – for better outcomes. Using technology will help to reduce errors and cut down on malpractice claims against physicians and institutions. Meanwhile, medical insurers will be happier because they won't have to spend money settling malpractice claims, and doctors will benefit by paying lower premiums.

HEALTHCARE IN THE UK

I'm proud to be an American, but I still think we can learn a thing or two from the British when it comes to Healthcare.

The National Health Service (NHS) is miles ahead of the US system, and they provide some interesting insights into how the future of healthcare could work in practice. In fact, in a recent comparison of the healthcare systems of eleven countries, the UK came out on top, beating out Canada, Australia, France and

Germany to take the top spot for the second year in a row. The US, meanwhile, was at the bottom.[41]

But the NHS is far from perfect. Funding has always been (and will continue to be) a problem, and the UK also has one of the highest rates of avoidable deaths in Western Europe. Kate Andrews of the Institute of Economic Affairs said that the NHS is "far from being the envy of the world", adding "it's not just low-income earners who receive poor care. The NHS's provision of care is equally poor for everybody, irrespective of income."

At the same time, rising life expectancies are slowing to a stop in the UK after more than 100 years of continuous progress.[42] Healthcare experts are divided about the cause, citing everything from government austerity to social factors such as education, employment, working conditions and poverty. It's also possible that we're approaching a natural limit to human life expectancy. After all, US doctors have already found some evidence to suggest that our longevity may be capped at around 115-years-old.[43]

The NHS has its problems. But it's also one of the closest things there is in the modern day to a value-based service, despite the fact that there's a long way to go. It could still benefit from interoperable data and a greater focus on prevention, but at the same time the biggest problem is that the organization itself is underfunded and struggling to meet unrealistic quotas. It's a step in the right direction, but we'll need to take more than a step if we want to make the future of healthcare a reality.

USHERING IN THE FUTURE

The current system is broken.

---

[41] See: http://bit.ly/nhsnumberone

[42] See: http://bit.ly/lifeexpectancyhalt

[43] See: http://bit.ly/limittolife

There's only one remedy for it: value-based healthcare. Instead of putting themselves first and focusing on the current, fee-for-service based model of healthcare, stakeholders need to usher in the future – a future in which value-based healthcare is the new norm.

Let's take a look at what that future will look like.

CHAPTER THREE: VALUE-BASED HEALTHCARE

"THE MOST EFFECTIVE MEDICINE IS THE NATURAL HEALING POWER THAT CAN BE FOUND WITHIN EACH AND EVERY ONE OF US."

- HIPPOCRATES

IF THE FUTURE OF HEALTHCARE is to become a reality, the industry needs to switch from its current fee-for-service model to a value-based model that takes a more holistic view of healthcare.

In the fee-for-service model, volume is king. Providers make more money if they carry out more tests and procedures, so it's no wonder that healthcare costs have skyrocketed in recent years. That's why efforts are already underway to reimburse providers based on the value (and not the volume) of the care they give. Medicare plans to shift 50% of its payments to these programs, which reward doctors for keeping patients healthy, by 2018.

This is a step in the right direction towards a value-based model that's all about quality. It's a little bit like the difference between purchasing luxury goods from a high-end store and opting instead for the cheapest option from the local Walmart.

In healthcare terms, this means that instead of every patient receiving bespoke treatment that aims to help them in the long-term, they typically get given the cheapest, least time consuming, most one-size-fits-all treatment that's available. But when it's your health at stake, shouldn't you demand – and expect – the very best?

The World Economic Forum put it best in their 2017 *Value in*

Healthcare report. "Value-based healthcare is a genuinely patient-centric way to design and manage health systems," it explains. "Compared to what health systems currently provide, it has the potential to deliver substantially improved health outcomes at significantly lower cost."[44]

While we wait for this shift to occur, we're stuck with the current fee-for-service model where the doctor/patient relationship is transactional, like the relationships we have with the cashiers at the grocery store. It's a lose/lose situation for every stakeholder, as we'll see later on in the book.

In this chapter, we'll spend some time looking at electronic health records (EHRs). These are subtly different to electronic medical records (EMRs), but the two terms are often used interchangeably and throughout this book, I use the term 'EHRs' to refer to both EHRs and EMRs.

When you're considering the difference between EHRs and EMRs, it helps to think about the terms 'health' and 'medical'. An EMR is a narrower view of a patient's medical history, while an EHR reports on the patient's overall health. EMRs are used by providers for diagnosis and treatment and aren't designed to be shared externally. EHRs are used to share a patient's information with authorized providers so that it can be accessed by specialists, labs, emergency rooms, pharmacies and more.

HEALTHCARE AT HOME

Most of what influences a patient's health occurs outside of a medical setting. For every five minutes that a patient spends in front of a doctor, they spend days or even weeks in the big, wide world. This is one of the reasons why randomized controlled trials (RCTs) are often less useful to clinicians than real-world/pragmatic trials.

---

[44] See: http://bit.ly/wefvalueinhealthcarereport

RCTs are specifically designed to eliminate bias and typically involve several groups of similar people being given different clinical interventions. The most basic example of this is some patients being given a new drug while others are given a placebo. Pragmatic clinical trials (PCTs) focus more on the correlation between treatments and outcomes in the real-world, as some treatments may not work as intended outside of the laboratory. Until new drugs and treatments are tested in a real-world setting, it's hard to tell whether they'll work as intended – and even whether there could be unexpected side-effects.

Insurance providers and fee payers have taken note of this. The emergence of mobile health technology (mHealth) is slowly but surely pushing us towards an era in which providers and researchers alike can get real world, real time data about patients' quality of life.

Unfortunately, providers are ill-equipped to take advantage of these mHealth technologies in our current healthcare system. Even if a patient is using wearable technology (such as a Fitbit or an Apple Watch), it's ultimately useless unless providers have time to analyze the data and implement the best course of action for each patient. Not every patient with hypertension or depression requires the same treatment plan. Technology and open dialogue between patient and provider has the potential to translate into lower costs and better outcomes for all.

It's estimated that, of the $6.5 trillion that's spent annually on global healthcare, up to 50% of it is wasted.[45] That adds up to a whopping $3,250,000,000,000 every year – vanished into thin air thanks to the inefficiency of our current healthcare system.

This could explain why Medicare and other private insurers are advocating for a switch to a new, value-based healthcare model that revolves around decreasing costs by keeping the patient healthy and out of the hospital.

Benjamin Franklin once said, "In this world, nothing can be said

---

[45] See: http://bit.ly/wefhealthcare

to be certain except death and taxes."[46] In the same spirit, I'd argue that our healthcare system's transition to a value-based model is inevitable. In fact, in a 2016 survey of healthcare executives and decision-makers, 94% of them admitted to being on the path to value-based care.[47]

The transition from a fee-for-service model to a value-based model won't happen overnight. Unfortunately, the alternative payment models for providers which are currently offered or being proposed by Medicare involve additional payments on top of fee-for-service. Like a heroin addict, we seem unable to go cold turkey and insist instead on just a little of our destructive drug fix. But this approach, like a bulldog that's toothless from birth, is doomed. If we truly want to live in a value-driven world, we need to be prepared to drop the fee-for-service model completely.

That said, I know from experience that not all stakeholders share this point of view. Humans are creatures of habit, so perhaps it's unsurprising that we're resistant to change. Change takes time, but ultimately the naysayers need to get on board the moving boat and sail with the tide while there's still time.

Resistance is futile because the value-based model is going to happen with or without them. The world is rapidly changing, and every day we wake up in a new environment. Asking 'why change the broken fee-for-service model?' reminds me of when Blockbuster was offered the chance to buy Netflix and turned it down.[48] It's worth noting here that Blockbuster's business model relied heavily on charging customers for late returns, while Netflix focused on adding value. Blockbuster didn't want to change their business model either, and look how that turned out.

---

[46] See: http://bit.ly/franklindeathtaxes

[47] See: http://bit.ly/adwaitgore

[48] See: http://bit.ly/blockbusternetflix

PAGING DR. WEED

While I was working on this book, I read the sad news that healthcare pioneer Dr. Larry Weed passed away at his home in Vermont, USA, aged 93.

Dr. Weed might sound like an advocate for medical marijuana, but his real claim to fame is his trailblazing work back in the 1950s. Charles Safran, Professor of Medicine at Harvard, said: "To describe his work as revolutionary almost understates it."[49]

Born on Boxing Day in 1923 in Troy, New York, Weed grew up to become a revolutionary whose work involved the creation of a new system to store patient data. This was before computing power became affordable – and before the machines themselves were small enough – for them to be a realistic option in a hospital setting. And because of that, it's a great example of how the future of healthcare was being ushered in long before artificial intelligence and big data were the big buzzwords that they are today.

Dr. Weed's system later became the first electronic health record (EHR) system, and you could argue that EHRs have more of an impact on modern healthcare than any individual drug – with the possible exception of the discovery of penicillin.

Before Dr. Weed, there was no way of measuring the effect of healthcare activities across a wider population. Every patient was treated in isolation and we had no way of monitoring long-term health trends. Weed's work is a precursor to the value-based healthcare system because the only way to make the shift is to base it on data.

Without Dr. Weed, the world is a darker place. But his legacy will live on as we continue to push for a holistic approach to healthcare.

In fact, progress is already being made. The Cleveland Clinic and CVS Health have formed a partnership that aims to improve patient

---

[49] See: http://bit.ly/drlarryweed

care in two states (Ohio and Florida) using data-sharing techniques and interoperability between different EHRs.[50] Hermann Stubbe MD, chairman of the Cleveland Clinic Florida Department of Family Medicine, explained that "the ability to share information, quality measures and protocols will reduce chronic disease and ultimately improve our patients' quality of life."

This lack of interoperability between different EHR providers is one of the biggest stumbling blocks that we'll need to overcome if we're to push forward into the future of healthcare. We're all indebted to Dr. Weed for his pioneering work, but if we don't allow different EHR systems to talk to each other then we're limiting their full potential – and not allowing Dr. Weed's legacy to live on in the way that it should do.

THE FUTURE OF EHRS

For the future of healthcare to be brought about, EHRs will need to be opened up and allowed to communicate with each other. It hasn't happened yet, but there's mounting pressure for records to be shared more freely, including with the patients they cover. This reminds me of the tense exchange between Epic Systems CEO Judy Faulkner and former Vice President Joe Biden. They were meeting to discuss the Cancer Moonshot, a White House project aimed at improving cancer research.

While discussing the possibility of giving patients access to their medical records, Faulkner reportedly said, "Why do you want your medical records? They're a thousand pages of which you understand ten." Biden responded with, "None of your business. If I need to, I'll find someone to explain them to me. And by the way, I understand a lot more than you think I do."[51] This exchange is made even more

---

[50] See: http://bit.ly/clevelandcvs

[51] See: http://bit.ly/noneofyourbusiness

poignant when you consider that they were discussing the 21st Century Cures Act, a recently-passed act which aims to bring patient records together in a "single, longitudinal format that's easy to understand."

Greg Simon, the president of the Biden Cancer Initiative, says that companies "want to make (and are required to make) only summaries available." He adds, "We've made billionaires of the executives of these companies. They've had fun. Now it's our turn. Let's get this thing changed." Simon also has interesting views on the ongoing debate about privacy when it comes to data sharing. "That's not coming from the patient community," he explains. "It's coming from the legal community and the consumer community."[52]

Companies like Epic (which dominates the EHR market in the US) are an obstacle for as long as they refuse to share data between systems. Why the hell should EHR companies decide what's relevant to the patient and what isn't? They think that the patient is incapable of understanding their own health, but they forget that we're all stakeholders in the system – and that if we can't take responsibility for our own health, no one can.

This whole scenario reminds me of *The Package*, an episode of *Seinfeld* in which Elaine is unable to receive medical treatment for a rash because of her reputation as a "difficult" patient. The episode originally aired on October 17th 1996, a couple of months after HIPAA was enacted. HIPAA is the Health Insurance Portability and Accountability Act of 1996, a piece of US legislation which is designed to offer data privacy and security provisions to safeguard medical information.

HIPAA is the big excuse that people give when it comes to data interoperability. Everyone has a right to data privacy, of course, but they also have a right to their data. Interoperability won't be a problem if the correct procedures are in place, which is why security is so important. The truth is that every time banks are hacked and it makes the headlines, everyone in the healthcare industry starts to

---

[52] See: http://bit.ly/epicfaulkner

worry about HIPAA. In my opinion, HIPAA itself is the problem. It does a great job of preventing the patient from accessing their information while the EHR companies and insurers get to sell the data to their clients.

The *Seinfeld* scene captures some of the problems that doctors fear will happen if they share their patients' data with them. Sure, perhaps there'll be an awkward moment when the patient reads what the notes say – but there's also every possibility that the notes are incorrect and the patient will be the one to point that out. Either way, it's time for HIPAA to catch up with the changing times. The fact that patients are being open with their data is fantastic as HIPAA doesn't prevent a patient from sharing their own data. And besides, HIPAA is over twenty years old. It was passed on the same day that Netscape 3.0 came out and the two of them have aged about as well as each other.

PERSONALIZED HEALTHCARE

I hold the view that in the future of healthcare, treatments will be tailored to match the unique needs of every patient.

Just as no two fingers on your hand are the same, neither are any two patients. Personalized healthcare is a sweeping umbrella term that includes patient behavior, communication, genetics, genomics and other biological information that helps to predict disease risks or to identify how a patient will respond to different treatment options.

An example of personalized healthcare is the inclusion of specific biomarkers – such as Lopoprotein (a) or apolipoprotein CD (APOC3) – which can help to better predict the risk of heart disease or stroke in certain individuals. These biomarkers can augment our current approach to risk assessment, which relies on factors such as age, menopausal status, diabetes, blood pressure or cholesterol levels.

I'm fortunate enough to have been involved with a startup in Palo Alto that developed the first – and as of now, the *only* – blood test that can help providers quickly, accurately and safely assess whether or not a patient's chest discomfort and other symptoms could be due to obstructive coronary artery disease. Without this blood test, the patients would be exposed to radiation during routine scans and, if they were female, would probably be referred to a cath lab only to be told, $25,000 later, that their symptoms were caused by heartburn and not an incoming heart attack. Notice that I haven't even mentioned the medical risks involved in undergoing cardiac catherization.

The molecular diagnostic algorithm we used in this particular case examines an individual patient's age, sex, and gene expression to calculate a score. Combining the score with other patient information provides clinicians with a more complete picture of the patient's status through the test's clinical validity in identifying patients who are unlikely to have obstructive CAD. This test was cheaper than both caths and nuclear stress tests and outperformed nuclear stress tests in ruling out obstructive coronary disease. Effectively, the algorithm was better at identifying people who might be at risk from heart problems than using radioactive material to check the size of the heart's chambers and how well it's pumping blood. It's a cheaper and safer solution, too.

The interesting thing here is that it's not just about diagnosis and treatment. The patient's entire encounter with the healthcare system needs to be personalized, and that includes improved communication between patient and provider.

Communication is highly important in healthcare, just as it is in other areas of life. In the current healthcare system, providers are overburdened, so tied up entering information into EHR systems and running from patient to patient that they don't have quality time to communicate with people. In the future, patients will expect this communication to take place regularly and via a medium that suits them.

Thankfully, technological advances will make this possible. Eric Topol captures this concept in his book, *The Patient will See You*

Now.[53] There is no "one size fits all" approach to communication. Seniors might prefer printed letters and face-to-face meetings while millennials will be much more comfortable with instant messaging, video conferencing and artificial intelligence.

Personalized healthcare and smart medicine will require a similar rethink. Providers will need to develop a deeper understanding of every patient on their roster, and the patients themselves will need to be empowered to play a part in their own treatment. This could include everything from wearing devices to track heart rates and exercise levels to monitoring themselves in their own homes. The FDA has already cleared some wearables that can monitor each of our vital signs, and taking advantage of this technology would reduce the amount of time that people spend in hospital. Less time in hospital means more efficiency, lower costs and a reduced rate of nosocomial (or "hospital-acquired") infections.

Meanwhile, the rapid growth and widespread adoption of mobile health (mHealth) technologies – which involve the use of mobile devices to monitor and educate consumers about preventative healthcare services – will give researchers, providers and patients access to objective, real world data about their health conditions. This will be invaluable when it's time to design treatment plans and rehabilitation programs.

Educating patients allows them to participate in their own healthcare, which we'll talk more about in chapter eight. This model of care is commonly known as shared decision making, and multiple studies have outlined its benefits.[54]

WHAT'S STOPPING US?

It's simple, but it's also complicated. The healthcare industry is

---

[53] See: http://bit.ly/patientseeyounow

[54] See: http://bit.ly/shareddecisionmakingpaper

difficult to change thanks to its entrenched culture and behavior, especially if you go up against the existing monetary models. Our current healthcare system is in such disarray that providers, hospitals and other stakeholders are so busy trying to make ends meet that there's no time left to think about the future.

Providers are often cited as one of the hurdles that's stopping a value-based system from being implemented, but the truth is that many doctors are pushing for this change to take place. According to Abhinav Shashank, cofounder and CEO of InnovAccer, physicians and their attitudes towards the value-based system can be classified into three groups:[55]

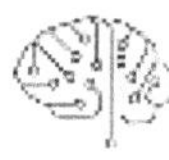

WILLING: Physicians who are likely to participate in a value-based system if they're given supporting tools such as interoperable EHRs and value-based technologies.

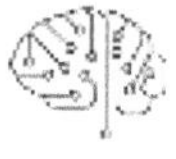

ON THE VERGE: These are the people who need to be won over. They're cautious but optimistic about value-based healthcare, but they also don't want to rock the boat. They have little access to tools and training but they may be turned into willing participants and value-based healthcare advocates given enough exposure.

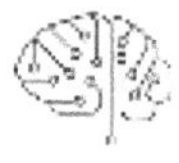

RESISTANT: These physicians are skeptical of and even actively against a value-based system. Even with appropriate incentives, they're unlikely to change their mind. Fortunately, in my experience at least, this mindset is more prevalent in older providers who are more jaded from their years of experience within the existing system.

But it's not just doctors who need to take responsibility. For a truly value-based model to succeed, we need to usher in the future

---

[55] See: http://bit.ly/abhinavshashank

from all angles, empowering patients with full transparency, using new technologies to improve their outcomes and shifting towards prevention instead of the cure. This shift will ensure that humans in the future will live longer and enjoy a better quality of life, instead of living a longer life that's marred with a myriad of debilitating chronic diseases.

At first glance, this seems to work against the financial interests of pharmaceutical companies. After all, they make their money by selling medication, and preventative medicine will certainly decrease the number of patients who otherwise would have been on therapies earlier on in life. That's why pharmaceutical companies will need to revamp their business models and start thinking beyond the pill.

In today's world, where value-based contracting deals between pharmaceutical companies and payers are on the rise, pharma needs to consider packaging pills with educational information and patient support services. These services will help patients with chronic illnesses to keep them under control with lifestyle changes, education and support, and this in turn will mean fewer visits to the doctor, fewer hospital stays and a better overall quality of life.

Outcome will be the name of the game. Healthcare companies that can't improve outcomes on a patient-by-patient basis will fall behind and have to settle for volume sales at low prices to survive. But patients will have greater choice and greater power, and it will fall to pharma companies to work with payers to secure access to these premium drugs with superior outcomes.

Perhaps this sounds naive of me, as one of the main hurdles in our healthcare system is affordability. I'll leave it to the politicians to sort that out – if they ever do – but in the meantime I'll have my dreams and live in utopia while I wait for healthcare policymakers to wake up from their phantasmagoric stupor.

MARKET POSITIONING

The healthcare industry is similar to many others in terms of how it operates. It's increasingly siloed, with companies specializing in one of the following three market positions:

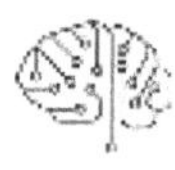

LEADER: These institutions are at the forefront of innovation, constantly engaged in research and development. Many of these leaders are government-sponsored not-for-profit organizations or independent think tanks within larger corporations. It's difficult for leaders to make money because of the associated costs of development, but when they're able to make a profit it puts them at the cutting edge of modern medicine.

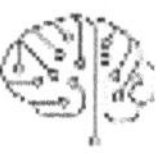

FAST FOLLOWER: These companies make their money by being second-to-market with new developments, allowing others to invest the cash up front and following suit as soon as they've seen a potential return. Many of the largest pharmaceutical companies are fast followers, and that's where the majority of the profit is under the current, fee-for-service system.

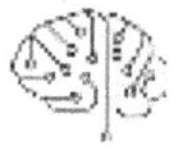

STRAGGLER: Stragglers are slow to market, often because of internal bureaucracy or heavy regulation. Most stragglers are either smaller companies under poor management or larger companies that are crippled by their size like an oil tanker that takes hours on end just to turn around. Stragglers usually struggle to operate at a profit and often place the least value on customers' welfare – because they're struggling to stay afloat and desperate to make some money.

If you want to find out more about how leaders, fast followers and stragglers will shape the healthcare industry, I recommend *Healthcare Disrupted* by Jeff Elton and Anne O'Riordan.[56] In it, they explain how in the volume-based fee-for-service era, more was better

[56] See: http://bit.ly/healthcaredisrupted

for most companies except for the stragglers – who were unable to gain traction due to poor management, lack of vision and uncompetitive products.

Now, though, we're moving towards a value-based future in which this three-tier classification is no longer appropriate. Consider autism spectrum disorder (ASD), an umbrella term for a group of developmental disorders which affect a person's social interactions, communication, interests and behavior. Autism is measured on a spectrum instead of under strict classifications, and we'll see something similar when it comes to our new healthcare system. The market will decide who survives, and this level of competition will bring healthcare into the 21st century, fostering innovation and creating a culture of continuous improvement.

MY VISION

We have a long way to go, but there's hope. There's a future out there that's brighter than anything you could imagine, and this book only scratches the surface. I can't wait for it to become a reality.

Patients won't have to go to the hospital. They'll be able to call their doctor on their phone or on Skype or Facetime. They can talk to a counsellor. They won't need to wait days, weeks or months for an appointment and healthcare will be affordable for all.

In our current system, we focus our time and resources on symptom relief instead of on disease prevention. Take diabetes. Five million people die of diabetes every year, mostly from Type 2 diabetes, which can be prevented through adjusting lifestyle factors such as losing weight, increasing exercise and improving nutrition. I agree with Jijo James, Chief Medical Officer at Johnson & Johnson Medical Devices, who believes in a new approach. Instead of diagnosing and treating, we'll prevent, intercept and cure[57], and

---

[57] See: http://bit.ly/jijojames

instead of focusing on prolonging life, we'll focus on improving its quality. Pharma's business model will need to change from "drug only" to "drug plus service", with services including patient education and support on disease prevention and management.

New technologies are pretty cool. I'm a geek when it comes to tech and its applications, and it's one of the things that got me interested in the future of healthcare to begin with. But technology is nothing unless it produces results, and that's what excites me the most.

Let's take a look at a few of those new technologies – and the results that they're able to generate.

CHAPTER FOUR: THE INTERNET OF THINGS

"THE INTERNET WILL DISAPPEAR. THERE WILL BE SO MANY IP ADDRESSES, SO MANY DEVICES, SENSORS, THINGS THAT YOU ARE WEARING, THINGS THAT YOU ARE INTERACTING WITH, THAT YOU WON'T EVEN SENSE IT. IT WILL BE PART OF YOUR PRESENCE ALL THE TIME. IMAGINE YOU WALK INTO A ROOM AND THE ROOM IS DYNAMIC. AND WITH YOUR PERMISSION AND ALL OF THAT, YOU ARE INTERACTING WITH THE THINGS GOING ON IN THE ROOM."

– ERIC SCHMIDT, GOOGLE CHAIRMAN

THE OBJECTS THAT WE INTERACT WITH are increasingly connected to the internet. In 2015, there were around 15.4 billion connected devices, which is two for every person on the planet. That figure is expected to rise to 30.7 billion in 2020 and 75.4 billion by 2025.[58]

But the important thing to remember about the internet of things is that it's not just about connected fridges that order more milk when you run out. In fact, the true potential of the internet of things is yet to be fully realized. We'll have access to more data than ever

---

[58] See: http://bit.ly/iotstatistics

before thanks to the plethora of connected devices, and we'll be able to use that information to improve our lives. Connected toothbrushes will help us to improve our dental hygiene. Connected fridges will help us to follow a healthy diet. Connected golf clubs will help us to improve our swings over time.

The internet of things, then, is all about connecting physical objects to the internet. But on top of that, it's also about what this internet connectivity will allow us to do in both developed and developing countries. Farmers could use IOT devices to monitor their crops and to maximize yields, helping to solve the hunger crisis in the countries that can't afford to create 30% of the world's garbage, like America can.[59]

Meanwhile, in the Western world, the internet of things will revolutionize many aspects of our lives, making them simultaneously more comfortable and more informed. In the future, your car could double up as a mobile office, fielding calls, scheduling meetings and even driving you from place to place.

---

[59] See: http://bit.ly/recyclablestats

WHERE IOT FITS

The next four chapters are dedicated to four key technologies which have the potential to revolutionize healthcare, of which IoT is just the first. Here's how they all fit together:

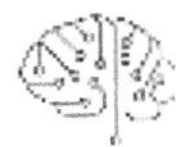

THE INTERNET OF THINGS: Allows us to capture individual data points for individual patients.

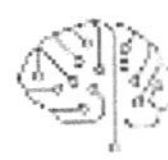

BIG DATA: Allows us to aggregate the individual data from IoT devices into a giant library of information.

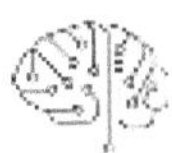

ARTIFICIAL INTELLIGENCE: Allows us to automate many tasks that previously would fall to humans but which are time consuming and often cost-prohibitive.

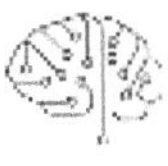

MACHINE LEARNING: Allows AI to further improve itself by "learning" through trial and error in the same way that a human learns.

AN ECOSYSTEM OF DEVICES

There's a war going on. Thanks to the open market and the relentless pace of innovation, thousands of different companies are creating IoT devices. This means that no one provider is taking over the market and that companies will need to create devices that can communicate with each other. If the devices are incompatible, it'll severely limit their capabilities and reduce their overall impact, which would be a shame for the future of healthcare, as well as the future of the world.

From a developer point of view, this causes something of a headache. Because there's no centralized system that they can tap into, they maintain all of their data within internal siloes that can't be accessed by other applications.

IOT IN HEALTHCARE

As we've seen, the internet of things has the potential to drastically change the way we live our lives, as well as the way we fight illnesses and diseases. In fact, developers are starting to put serious thought into the healthcare market, and not just because it's worth billions of dollars. They know that their devices could increase not only our average lifespans but also the quality of life that we enjoy along the way.

Unfortunately, it can be difficult to make a dent. Changing the existing healthcare system takes time and lawmakers are still struggling to keep up with the pace of innovation. Creating internet-connected technology is actually easier than bringing it into the workflow of today's systems.

And there's another problem, one which healthcare professionals are all too aware of but which startups and their developers keep forgetting. The patients who cost the most money – and who could therefore most benefit from internet-connected technologies like remote monitoring – have very little interest in being monitored.[60] Many of them are elderly, nearing the end of their life or suffering from debilitating diseases, and remote monitoring may seem like unnecessary extra hassle. The people who *want* to be monitored are those who least need it.

HOW DEVICES 'HOOK' YOU

Of all of the IoT devices on the market, none is more obvious – or as well-known – as the Amazon Echo. This IoT device is basically a hub for Amazon's Alexa, their AI-based virtual assistant, but both the device and the assistant are designed to "connect the user's

---

[60] See: http://bit.ly/iatmonitoring

problem with the company's product with enough frequency to form a habit" according to Nir Eyal, the author of *Hooked: How to Build Habit-Forming Products.*[61] In his book, Eyal outlines how IoT devices like the Echo hook their users:

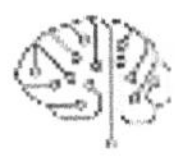

TRIGGER: These prompt us to take an action and tell us what to do next. Smartphone apps might attract your attention with a push notification, but IoT devices often require us to trigger ourselves. So with the Echo, a trigger might be our curiosity when we want to know the answer to a question.

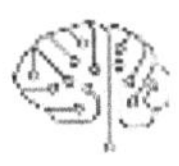

ACTION: A trigger is useless unless it pushes someone to take an action. People take an action – such as asking Alexa a question – after the trigger pushes them to seek a reward. The easier it is to take an action, the more likely people are to do it.

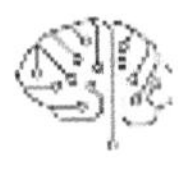

REWARD: This is where the user gets what they came for, which is usually a relief for the "itch" that caused the internal trigger. In the Echo's case, the reward could be information in the case of a search or confirmation when you're setting a reminder.

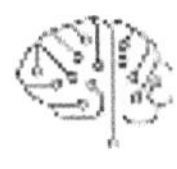

INVESTMENT: Eyal explains that the investment stage occurs when users "[put] something into the product to improve it with use." With Netflix, its recommendations get better and better the more you rate content. Similarly, Amazon's Alexa can learn more about you to provide more accurate information. This phase is increasingly supported by machine learning to make the customer experience as bespoke as possible, no matter what device or application you're using.

---

[61] See: http://bit.ly/gettinghooked

You might be wondering why I've included this information in a book about the future of healthcare. The reason behind it is pretty simple: If future healthcare services and devices are to be successful, they'll need to hook users in much the same way that Eyal described.

After all, Eyal's book is about building habit-forming products, and habits aren't always bad for us. If the healthcare industry can learn from the technology industry and start encouraging positive health habits such as exercising regularly and eating healthily, it can start to save itself huge amounts of money through preventative healthcare. For this to happen, there needs to be a shift in mindset. Luckily, I firmly believe that we're already on our way towards a more holistic future in which technology will be used to improve population health as a whole.

THE ROLE OF DISRUPTION

Disruption is inevitable. Just ask Alberto Brea, Executive Director of Engagement Planning at Ogilvy, who says: "Netflix did not kill Blockbuster. Ridiculous late fees did. Uber did not kill the taxi business. Limited taxi access and fare control did. Apple did not kill the music industry. Being forced to buy full-length albums did. Amazon did not kill other retailers. Bad customer service and experience did. Airbnb isn't killing the hotel industry. Limited availability and pricing options are. Technology by itself is not the real disruptor. Not being customer centric is the biggest threat to any business."

The same can be said of healthcare and its adoption of machines and new technology. When something is broken, someone will eventually come along and fix it, and that's how the future of healthcare will be ushered in.

According to a report from MarketResearch.com, the healthcare segment of the internet of things market will hit $117 billion by 2020.[62] That might sound like a lot of money, but that's nothing compared to what's coming. As companies start to realize the huge potential of the internet-connected healthcare market, budgets will be raised, new standards will be agreed on and the industry as a whole will slowly start to follow the path of least resistance. The inevitable path into the future of healthcare.

THE INTERNET OF HEALTH THINGS

The Internet of Health Things (IoHT) is the term that's used to refer specifically to IoT-connected health devices, and the industry is already gigantic. One study from eMarketer found that the IoHT market will be worth $163 billion by 2020, with a compound annual growth rate (CAGR) of 38.1% between 2015 and 2020.[63]

In fact, the huge amount of growth in the area is set to push the healthcare sector to the top of the list of industries that are leading the way with IoT development. A different report from Accenture also noted that the insurance industry will witness huge disruption from artificial intelligence, with 32% of insurers saying that their company will be "completely transformed" by AI within the next three years.

This all goes to show that the future of healthcare might not be as far away as we thought. It's not just something that we can expect to see in our lifetime – it's already happening, and the pace of innovation will only continue to accelerate.

The next generation of IoHT devices will likely focus on three key areas: remote patient monitoring, wellness and prevention and operations. Their benefits range from cost savings and revenue gains to an improved patient experience, reinforcing the fact that the

---

[62] See: http://bit.ly/117billion

[63] See: http://bit.ly/IoHTmarket

future of healthcare is about how services are delivered, as well as what the services are to begin with.

AI AND THE INTERNET OF THINGS

Internet-connected devices are only part of the story. After all, with billions of different devices hooked up to the internet and countless different data points, we'll need a big "brain" to process it all.

That's where artificial intelligence comes in. The internet of things collects the data and artificial intelligence crunches the numbers to make sense of it. The idea is to identify patterns and to use data to make more informed decisions, opening up the ability for proactive intervention before diseases and illnesses become a bigger problem. AI is perfect for the job of processing huge amounts of data, and it can also take advantage of intelligent automation and bespoke recommendations to usher in an era of personalized medicine.

Unfortunately, the development of artificial intelligence is lagging behind the development of the internet of things[64], which may mean we face a disappointing wait before we're able to truly understand the data. However, every analysis requires some data to begin with, and if IOT devices allow us to collect data for several years before AI technology has caught up with it, that might not be a problem. It'll just mean that we'll have a deeper pool of data with which to train software.

Now, it might sound as though this high-tech combination of AI and IoT is only applicable in futuristic hospital settings in a world that's managed by robots, but nothing could be further from the truth. In fact, it's already being used in both developed and developing countries.

---

[64] See: http://bit.ly/whyiotneedsai

In India, citizens have little access to doctors but plenty of internet connectivity. It's also worth noting that the nation had a late start and so it was able to sidestep much of the early development of the internet. Indians skipped desktop computers entirely and became widespread adopters of the mobile internet, with 80% of Indian internet access occurring through a mobile device.[65] Most of the innovations and technological breakthroughs in medicine are limited to the big city hospitals, which do little to help the rural populace that makes up 72% of the country. That's why a greater emphasis is being placed on preventative care to diagnose potential problems ahead of time and to head them off at the pass.

The problem in rural India is that of how to deploy diagnostic tools to the front line. There's a shortage of specialists in rural health centers[66] and so empowering people to take charge of their own health will be a huge driver of change in the future. And when you consider that these rural citizens are already using mobile devices to connect to the internet, the next big step for Indian healthcare will be the development of new applications for existing mobile technology.

Technology is no substitute for a medical professional, but it can help them to reach out to patients in disparate and hard-to-access areas. Remote consultations, chronic disease management and overall prevention becomes feasible – but only when IoT and AI are used together to act in patients' best interests.

---

[65] See: http://bit.ly/indiastats

[66] See: http://bit.ly/indiaai

CHAPTER FIVE: BIG DATA

"BIG DATA REALLY ISN'T THE END UNTO ITSELF. IT'S ACTUALLY BIG INSIGHTS FROM BIG DATA. IT'S THROWING AWAY 99.999% OF THAT DATA TO FIND THINGS THAT ARE ACTIONABLE."

- BOB BORCHERS, CHIEF MARKETING OFFICER AT DOLBY LABORATORIES

BIG DATA has the potential to change every aspect of our lives. Google's Eric Schmidt famously said that we create as much data every two days as we did from the dawn of time until 2003.[67] The Economist, meanwhile, suggested that data has overtaken oil as the world's most valuable resource.[68]

Too much information can be a bad thing. In fact, a team of neuroscientists from Canada's University of Toronto published a paper which hypothesized that our brains purposefully forget information to help us to live our lives.

The same concept applies to the complicated world of artificial intelligence. There's a phenomenon called over-fitting in which a machine stores so much information that it's unable to behave intelligently. Blake Richards, one of the co-authors of the paper, says that understanding how our brains decide what to keep and forget

---

[67] See: http://bit.ly/schmidtdatafact

[68] See: http://bit.ly/nolongeroil

will help us to create better AI systems that can more tightly integrate with the world.[69]

AI and machine learning technologies have always faced problems when making the leap from the lab to reality. A prime example of this is Microsoft's Tay, an AI chatbot that was recalled within 24 hours after Twitters users taught it to be racist and misogynistic.[70]

Nevertheless, if the future of healthcare is to be ushered in then AI systems are going to need to work in the real world – and not just in hospitals and research laboratories. Big data is what will help us to do that.

Economists are already calling data "the new oil", and it's certainly true that the companies with the biggest – and best – collection of data will be the ones who earn the most money. But when it comes to healthcare in particular, a question is raised: Does this data come at the expense of individuals? And will people be happy to give up their medical histories to corporations that they might not be able to control?

One of the challenges for big data in healthcare will be the interoperability of systems and the drive to secure patient consent to securely store their information. To do this, we need to introduce new technologies carefully, pre-empting potential issues by embracing these difficult conversations now instead of waiting until they become a problem.

Still, the major players in the industry are already making moves. Amazon, Microsoft, Apple and Google are all showing signs of collecting and analyzing big data for the purposes of healthcare, and Facebook's Mark Zuckerberg has already hinted at an entry into the market. Alongside wife Priscilla Chan, he donated $10 million to support UC San Francisco's Institute for Computational Health Science in their goal of advancing health using big data.[71]

---

[69] See: http://bit.ly/memoryforgets

[70] See: http://bit.ly/racisttay

[71] See: http://bit.ly/zuckdonation

The good news is that if this data is made available to researchers, we could bring our current system into the 21st century. Take the study published by Chris Danforth (University of Vermont) and Andrew Reece (Harvard University) in the EPJ Data Science Journal. The researchers developed an algorithm that scanned Instagram posts and could accurately identify depression in 70% of their study's participants. This contrasts with the 42% success rate for traditional doctors.[72]

BIG DATA IN HEALTHCARE

Much of big data's potential for the healthcare industry has already been covered elsewhere. Over the last few years, we've come along in leaps and bounds, but there are still several key ways in which big data could be put to better use, as Ryan Ayers explained in an article for Dataconomy:[73]

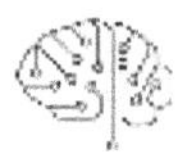

PREVENTING MEDICATION ERRORS: At least 1.5 million preventable adverse drug effects (ADEs) happen every year in the United States alone, causing up to 7,000 deaths a year.[74] Big data can be used to better understand the interactions between different medications and to dramatically reduce the number of ADE events.

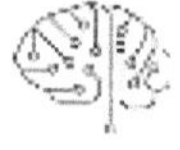

IDENTIFYING HIGH RISK PATIENTS: By combining big data with predictive analytics, it's possible for hospitals to reduce the number of visits to the emergency room by identifying high-risk patients and

---

[72] See: http://bit.ly/instagramdepression
[73] See: http://bit.ly/bigdataforhealthcare
[74] See: http://bit.ly/adesstat

offering them more personalized healthcare. However, more data needs to be gathered before these systems can be truly effective.

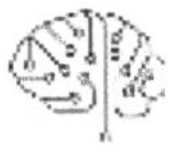

REDUCING COSTS AND WAIT TIMES: This also relies on combining big data with predictive analytics, this time to base staffing decisions on expected admission rates. This would enable hospitals to cut down on wastage and to adapt their services to meet demand, improving patient outcomes while simultaneously saving money.

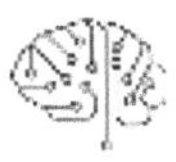

PREVENTING SECURITY BREACHES: The healthcare industry is 200% more likely to experience a data breach than other industries[75] because its vast amounts of personal data are valuable to hackers and fraudsters. But while big data can attract attackers, it can also be used against them – as is the case for the Centers for Medicare and Medicaid Services, which prevented over $200 million in fraud in a single year using big data analytics.[76]

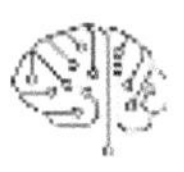

ENHANCING PATIENT OUTCOMES: This is perhaps the most important of the five points when it comes to the future of healthcare. Ayers notes that "consumer interest in devices that monitor steps taken, hours slept, heart rate and other data on a daily basis shows that introducing these devices as a physician aid could help improve patient engagement and outcomes." New tech and the big data that it creates will revolutionize the way that we look at healthcare, putting more power in the hands of patients. That can only be a good thing.

---

[75] See: http://bit.ly/securitybreaches

[76] See: http://bit.ly/bigdatainfraud

PREVENTATIVE CARE

The 1990s comedy-drama *Patch Adams* might seem like an unusual place to find inspiration for the future of healthcare, but Robin Williams' speech at the end of the movie is a true masterclass.[77] Just a small section of it is shown below:

Every human being has an impact on another. Why don't we want that in a patient doctor relationship? A doctor's mission should be not just to prevent death but also to improve the quality of life. That's why, you treat a disease, you win you lose. You treat a person, I guarantee you win no matter what the outcome.

That's why preventative care is so important. We need to make sure that our quality of life is kept from deteriorating – by treating obesity before the patient develops diabetes, for example. By having regular checkups, screenings and immunizations, we can stop bad things from happening instead of having to fix them later.

In an article called *Why Technology Will Transform Healthcare,* investment company Baillie Gifford introduced Doxcom, a company which makes a sensor that attaches to your skin and monitors blood-glucose levels in real-time and then uploads the information to the internet. Baillie Gifford explains, "Dexcom's products are primarily used for the diabetic market, and insulin-taking diabetics can dose directly off the information they provide. This massively improves their compliance, limits hospitalization and in theory gives much better long-term outcomes. But just think how a technology like this could evolve. At the moment, it's mainly used for type 1 diabetes but the much larger opportunity is for type 2 or 'lifestyle-induced' diabetes, which is far more common."

Sounds good, right? Here's how Baillie Gifford pictures the fight against type 2 diabetes: "Imagine a device where this information is sent to your phone which has a global positioning system chip in it. This chip will be able to determine how far you have walked, the

---

[77] See: http://bit.ly/patchadamstranscript

amount of calories you've used up, that it's lunchtime and that you've just walked into a restaurant. It will then be able to pull up the restaurant's menu from the internet and based on your unique profile it will suggest something to eat."

The technology here is nothing new. When he was working on the book, my editor got into the spirit of things by buying a Fitbit Charge HR, which measures the wearer's heart rate, as well as the steps they take and the sleep they get. The Fitbit also integrates with MyFitnessPal, which can be used to track calories, and the apps automatically allow you to eat more if you earn it through exercise. It's not much of a stretch to imagine a third app that could integrate with MyFitnessPal and Fitbit and make nearby food recommendations. And it's all made possible thanks to the power of big data.

This trend was highlighted in an article in The Economist which explained, "More data will not only identify those drugs that do not work. Digital healthcare will also give rise to new services that might involve taking no drugs at all. An app or a wearable device that persuades people to walk a certain distance every day would be far cheaper for insurers and governments to provide than years of visits to doctors, hospitals and drugs. Although Fitbits are frequently derided for ending up in the back of a drawer, people can be motivated to get off the sofa. Players of Pokémon Go have collectively walked nearly nine billion kilometers since the smartphone game was released last year."[78]

Big data has a huge role to play in the future of preventative healthcare. It can predict illnesses, cure diseases, improve quality of life and stop unnecessary deaths. This is my philosophy, and new technologies will help to get us there. That's how we'll drive down costs and benefit every stakeholder across the industry.

---

[78] See: http://bit.ly/digitalhealthrevolution

CHAPTER SIX: ARTIFICIAL INTELLIGENCE

"BY FAR THE GREATEST DANGER OF ARTIFICIAL INTELLIGENCE IS THAT PEOPLE CONCLUDE TOO EARLY THAT THEY UNDERSTAND IT."

- ELIEZER YUDKOWSKY, AI RESEARCHER AND AUTHOR

IT'S TIME TO TAKE A LOOK at artificial intelligence, which is advancing so rapidly it's hard to keep track of it. Just ten years ago, Akinator[79] – the web genie – was hailed as a triumph of artificial intelligence technology. The bot, which was developed by three French programmers, pitted human players against artificial intelligence in a game of Twenty Questions.

But in the decade since, artificial intelligence software has evolved and matured. In 2011, IBM's Watson competed against two former *Jeopardy!* winners and picked up the first prize of $1 million. By 2013, it was helping to treat lung cancer patients at Memorial Sloan Kettering Cancer Center in New York City. In 2014, a computer program called Eugene Goostman, which simulates a 13-year-old Ukrainian boy, became the first piece of AI software to pass the Turing test.[80] Named after computing pioneer Alan Turing, a piece of software passes the Turing test if a human being can't

---

[79] See: http://bit.ly/playakinator

[80] See: http://bit.ly/aituringtest

distinguish the machine from another human being using questions that are posed to both humans and machines. Later, in 2017, Google's DeepMind AI defeated the world's number one Go player, despite many people claiming it would be impossible for a computer to ever beat a human.[81] There are more potential Go moves than atoms in the universe, so no computer could be powerful enough to use brute force to decide upon a move. The same AI software was also used by the UK's National Health Service to develop apps and other tools for diagnosis purposes.

Deep Blue's defeat of Kasparov led an Indian-American AI programmer to invent a game called Arimaa in 2003. Arimaa is specifically designed to be played with a standard chess set and to be difficult for computers while being easy to learn and fun to play for humans. I think that says a lot about our attitude towards AI – as soon as it's about to beat us at something, we come up with something else that it will struggle with.

I recently read a fantastic article by Dr. Isaac Kohane[82] in which he discussed the care he provided for his elderly mother, who'd been hospitalized twice due to a complication from her heart dysfunction. Kohane realized that he could use internet-connected scales to remotely monitor his mother's weight, and that she could take an extra dose of Lasix to redress the balance before she needed to return to the hospital. Kohane developed an algorithm that helped to manage her health and called his mother regularly to remind her to weigh herself until it became second nature.

The interesting thing about Kohane's story is not that it's a fantastic example of personalized healthcare, although it is, but that much of this activity could be automated by a piece of artificial intelligence software and rolled out en masse to other people like his grandmother.

---

[81] See: http://bit.ly/googlebeatsgo

[82] See: http://bit.ly/futureofai

THE TYPES OF AI

Hypothetically speaking, there are three different types of artificial intelligence. At the moment, we're only working with AI at its most basic, but as the technology continues to mature, it's only going to grow to be more and more powerful:

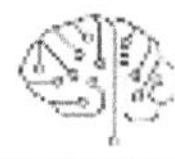

AI: Basic artificial intelligence does exactly what it says on the tin. It's essentially a piece of software which simulates intelligence without fully duplicating the complex processes of the human brain.

GENERAL AI: This is the next step of artificial intelligence in which AI software will be able to 'think' like a human being. At this stage, AI will no longer be a simulation for human intelligence but will be almost indistinguishable from the real thing.

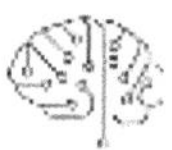

SUPERINTELLIGENT AI: This is the third hypothetical stage of AI's evolution in which it becomes infinitely more intelligent than a human. The only way to achieve this will be to use AI systems to develop more advanced AI systems and would mark a turning point in history at which computers were able to design themselves – which sounds like the premise of a post-apocalyptic sci-fi novel. Fortunately, we're a long way off needing to deal with the complex moral issues that such software would ask us to consider.

THE CHINESE FOCUS

Let's face it – tech (and business as a whole) in the United States is increasingly under threat from Chinese competitors. With growing global influence and more native Chinese speakers than native speakers of any other language, not to mention the world's largest population, China is a global powerhouse just waiting to happen.[83]

The country has already laid out plans to become the world leader in the artificial intelligence field by 2025[84], increasing government spending on AI to $22 billion in the next five years and up to $60 billion per year by 2025.

AI in the Chinese medical sector is expected to be worth 1 trillion yuan ($147 billion) within the next 20 years, and around 150 Chinese AI companies are already involved in the medical sector.[85] There's also a growing trend for people to treat themselves, with Fan Wei, director of Baidu's big data lab, explaining, "Only 4.8% of the urban population go to see a doctor when they feel sick. Meanwhile, up to 89% of the online population searches the internet for medical knowledge. Baidu Zhidao, the question and answer section of Baidu, receives 10 million medical inquiries every day. So the market for AI in the healthcare sector is huge."

Still, with Microsoft, Apple, Google, Amazon and IBM all in the game, the Chinese invaders will face tough competition – which is good for all-round innovation and the advancement of new technologies. Microsoft in particular has publically stated in its annual SEC report that "[its] strategy is to build best-in-class platforms and productivity services for an intelligent cloud and an intelligent edge infused with artificial intelligence ("AI")."[86]

The result will be the 21st century's equivalent of the space race. Only time will tell who'll be the first to touch down on the surface of the future of healthcare.

---

[83] See: http://bit.ly/thefutureoflanguage

[84] See: http://bit.ly/chinaai

[85] See: http://bit.ly/aitransform

[86] See: http://bit.ly/microsoftai

A SLOW GROWER?

AI is notoriously difficult. According to Robin Hanson in OpenMind[87], "When individual AI researchers have gone out of their way to make public estimates of the overall future rate of progress in AI research, averaged over all of the subfields of AI, their median estimate has been that human-level abilities will be achieved in about thirty years. This thirty-year estimate has stayed constant for over five decades, and by now we can say that the first twenty years of such estimates were quite wrong."

Artificial intelligence theorist Eliezer Yodkowsky highlights another of the issues we face: "By far the greatest danger of artificial intelligence is that people conclude too early that they understand it." That's part of the reason why it's so slow to be adopted by healthcare companies.

But despite all of this, the AI healthcare market is seeing strong growth that will continue for the years to come. In 2016, it was valued at $667 million, but it'll be worth $8 billion by 2022. Meanwhile, the healthcare IoT market is expected to rise from $22.5 billion in 2016 to $72.02 billion by 2021.

Put simply, data is knowledge, but that knowledge requires artificial intelligence if we ever hope to process it. Meanwhile, more and more devices are storing more and more information, which unlocks a huge amount of potential for AI. It could be used to determine the prevalence of different illnesses in different parts of the country, identify outbreaks in their early stages and even challenge established norms using data to back its arguments.

---

[87] See: http://bit.ly/humanlegacies

AI MISBEHAVING

Artificial intelligence doesn't always behave itself, and machine learning is part of the problem. Bots can be tricked and taught to swear, and they can even go haywire and do something totally unexpected, like when Facebook's artificial intelligence bots started to develop their own language. The sentences seemed like gibberish, but scientists with Facebook AI Research (FAIR) discovered that they comprised of a coded type of language that made sense to the bots but not to humans.[88] Facebook ended up killing the program and taking the bots offline – not because they were plotting to take over the world but because the whole point had been to create bots that were capable of talking to humans.

Another great example of rebellious AI came up in an article by Jamie Condliffe in MIT's Technology Review.[89] Condliffe explained how AI can be tricked into mishearing what a recording says or what is shown in an image, and it can be done in a way that's indistinguishable to humans. It might not sound too threatening at the moment, but imagine a hacker convincing self-driving cars to see fake traffic or ordering a virtual assistant to make spurious purchases. We can only imagine how much damage something similar could do to a health system that was insufficiently secured.

Luckily for us, these types of mishaps are due to programming errors and not due to an inherent flaw with artificial intelligence, and any AI technology that received sufficiently mainstream adoption for it to cause a problem would also have been thoroughly tested in a real-world environment.

Remember, the use of technology is nothing new. X-ray machines are far more dangerous than artificial intelligence, and yet they're a routine piece of kit in the hospital environment. It won't be long until AI follows in the X-ray's footsteps.

---

[88] See: http://bit.ly/facebookbotsrebel

[89] See: http://bit.ly/aiinmit

WHAT AI CAN DO

The potential of artificial intelligence is almost limitless, and the technology is far from unique to healthcare. It can be used to design logos[90] and create British place names[91] or to tap into Wi-Fi networks to determine our emotional state.[92] This latter technique works like an electrocardiogram (ECG) without requiring leads to be attached to the person. Instead, it uses the same technology that's inside routers to bounce RF signals off people and uses machine learning to identify a person's emotions based upon their heartbeat. In the future, the same setup could figure out that you're stressed and play relaxing music without you even knowing that you needed it. It could also tell when people are at risk of a heart attack.

AI can even sleep. Google's DeepMind is now so "human" that it's learned to mimic "experience replay" by "storing a subset of training data that it reviews 'offline', allowing it to learn anew from successes or failures that occurred in the past."[93] Like human beings, DeepMind will be able to think something through while it's asleep, with TheNextWeb explaining, "It doesn't have to be working on a problem to solve it. It'll fail at something, go offline, and then be able to succeed at that task once it's back online."

While I was researching this book, I kept my eyes peeled for different uses of AI that are already out there and active in the big, wide world. You might be surprised at just how many there are. Marketers are using AI to analyze social media activity and to plan out marketing campaigns.[94] Google is using AI to remove extremist content from YouTube because it's faster and more accurate.[95] AI

---

[90] See: http://bit.ly/ailogodesign
[91] See: http://bit.ly/boredprogrammer
[92] See: http://bit.ly/aiemotions
[93] See: http://bit.ly/sleepingai
[94] See: http://bit.ly/aiformarketing
[95] See: http://bit.ly/aijoinsthefight

technology is also used for voice assistants like Alexa, facial recognition on smartphone apps and automobile safety and autonomous car systems.[96]

The truth is that you're already using artificial intelligence whether you know it or not. When you upload a photo to Facebook, it uses artificial intelligence to suggest the name of the person who's shown. When your credit card is used, AI tries to identify patterns that are out of the ordinary to combat fraud. If you turn on automatic subtitles on YouTube, that's provided by artificial intelligence and machine learning. When you make a search on Google, AI is used as a factor to determine the results that you see. And, of course, Netflix provides its recommendations based on an AI system that predicts what users might be interested in based on their previous viewing habits, as well as from the viewing habits of other users.

These real-life uses of artificial intelligence are all well and good, but the technology's true potential won't be witnessed until it's deployed in revolutionary new ways to improve the outcomes of patients. Just one example is that of the emerging field of computational psychiatry.[97]

Borderline personality disorder is characterized by an unstable sense of self and unstable emotions. Sufferers are more likely to harm themselves than non-sufferers, and 10% of them eventually commit suicide. There's no known cause, but it is known that genetic and environmental factors play a role. Computational psychiatry aims to use big data and artificial intelligence to allow us to improve both diagnosis and overall understanding of the condition. People with borderline personality disorder tend to use language in certain unusual ways, and applying computational psychology to the field could allow us to develop a firmer understanding of how and why this happens.

---

[96] See: http://bit.ly/needtoknowai

[97] See: http://bit.ly/computationalpsychology

Venture capitalist and Sun Microsystems cofounder Vinod Khosla believes that artificial intelligence will eventually replace human oncologists.[98] Indeed, he went so far as to say, "I can't imagine why a human oncologist would add value, given the amount of data in oncology. They can't possibly comprehend all of the things that are possible." He says that AI has become the new poster-child for the scientific community, adding, "If I go to MIT and give a talk on AI, I can fill any hall instantly. If I want to talk about energy, I can't fill the halls."

ARTIFICIAL INTELLIGENCE IN HEALTHCARE

Artificial intelligence is particularly well suited to personalized healthcare because AI software can run calculations in bulk and use machine learning – which we'll discuss in chapter seven – to improve itself over time. The human brain can only do so much, and machines are often more efficient at recognizing patterns that are otherwise buried in huge amounts of data.

According to some interesting new research from New York University's Langone Medical Center, AI-based virtual assistants are getting smarter and smarter and may one day be able to save lives or to identify diseases earlier than ever. It essentially relies on using short voice clips to diagnose a variety of diseases and conditions.

Paul Armstrong, the author of *Disruptive Technologies*, explains, "Using complex algorithms and machine learning, the researchers hope to find vocal patterns that might signal illness and more complex disorders via a five-year study."[99] He adds that the science is complex but that "it's not hard to see how this sort of leap forward could be adopted quickly" and praises its potential to identify post-traumatic stress disorder and heart disease. It's easy to see how,

---

[98] See: http://bit.ly/vinodkhosla

[99] See: http://bit.ly/alexadiagnosis

theoretically at least, a virtual assistant could learn to spot signs of depression or anxiety in its user based on what it knows about that particular person.

Personalized healthcare is important because no two patients are the same. Let's say two people are in an emergency room with chest pain. One of them is a 26-year-old carpenter who fell ill while working on a building site while the other is a 72-year-old man with a history of strokes and heart problems. These factors need to be considered, and the two different patients have drastically different needs when it comes to treatment and aftercare.

That's where artificial intelligence comes in. When used correctly, it can help medical staff to make a diagnosis, tracking patients throughout their lives and using existing data to identify treatment options based on their medical history. It could also compare individual patients with their peers – so the 72-year-old could be compared to other 72-year-old men with heart problems and with similar socioeconomic, occupational, geographic, ancestral and genomic backgrounds to identify what worked in other cases, as well as what didn't.

Unfortunately, this kind of analysis will only work if the data is available. Historically, patient data has been stored offline or in internal systems that aren't shared with the wider world. It's there, but only if you're looking for it. But that's already starting to change, thanks to increasing cooperation and globalization and new devices like the NVIDIA DGX-1[100], the world's first deep learning supercomputer. Specifically designed for artificial intelligence, the futuristic machine can use demographic and lifestyle factors to predict the response that individual cancer patients are likely to have when given different types of treatment.

Back in 2016, Dr. James Brink and Dr. Mark Michalski (along with their team at Massachusetts General Hospital's Center for Clinical Data Science) were lucky enough to represent the first medical institute in the world and one of the first five research

---

[100] See: http://bit.ly/nvidiadgx1

institutions of any kind to receive the NVIDIA DGX-1.[101] It made its debut at the Hospital's historic Ether Dome, which is where the first public demonstration of surgery under anesthesia took place in 1846. The anticipation in the dome was palpable and looking back on it, it reminds me of the cold chills of excitement that ran down my spine as I listened to Dr. Michaelski talk about the future of AI in healthcare during a standing room only session at the World Medical Innovation Forum in 2017.

Of course, these revolutionary new machines are expensive when they first hit the market, which puts them out of reach of the majority of healthcare institutions. Fortunately, the cost of new technologies tends to fade over time, so we can look forward to a future in which powerful AI devices are as commonplace as defibrillators.

Meanwhile, we can already see the potential for AI when it comes to medical treatment. IBM's Watson already has a specialized program for oncologists which analyzes structured and unstructured data from clinical notes and reports before combining it with the patient's file to identify potential treatment plans. The process is overseen by a professional, of course, but Watson is able to do much of the legwork so that physicians have more time to dedicate to other duties.

In one recent study, Watson created a treatment plan for a brain cancer patient in just ten minutes, while doctors would have taken 160 hours.[102] Writing for IEEE Spectrum, senior associate editor Eliza Strickland explained: "Watson's key feature is its natural language processing abilities. This means Watson for Genomics can go through the 23 million journal articles currently in the medical literature, government listings of clinical trials and other existing data sources without requiring someone to reformat the information and make it digestible."

Strickland also notes that Watson's analysis wasn't perfect and

---

[101] See: http://bit.ly/nvidiadgx1blog

[102] See: http://bit.ly/watson10mins

that there's plenty of room for physicians and AI tech to work together. Rober Darnell, director of the NYGC and a lead researcher on the study, added, "Watson provided annotation that made the analysis faster. Given that each team [i.e. humans versus Watson] addressed different issues, this comparison is apples to oranges."

This highlights one of the main areas in which artificial intelligence is likely to be most successful. Medical practitioners are perpetually short on time, and AI software will help to free them up by taking on time consuming and repetitive tasks, like sieving through scans to detect anomalies or collating information on medical conditions to help patients to come to terms with a diagnosis.

Meanwhile, AI tools will be deployed at a genetic level to identify patterns across huge sets of data to spot mutations and to try to determine new genetic signposts that could be used to diagnose diseases. Deep Genomics[103] is already combining machine learning and experimental biology, but it's likely to become more prevalent in the future, especially once entrepreneurs and investors start to see the technology's potential. After all, if these tools are able to detect cancer or vascular diseases at an early stage, they'll significantly improve patients' chances of survival.

Prevention is better than the cure, but AI software and machine learning will help with both. In fact, AI software is already helping to develop new pharmaceuticals, with Atomwise using supercomputers to discover two potential Ebola treatments in less than a day. Alexander Levy, the company's COO, said, "If we can fight back deadly viruses months or years faster, that represents tens of thousands of lives. Imagine how many people might survive the next pandemic because a technology like Atomwise exists."[104]

---

[103] See: http://bit.ly/deepgenomics

[104] See: http://bit.ly/newebola

IMAGINING THE FUTURE

It's not just oncologists that could be at risk of automation. After all, writers have a 33% chance of losing their jobs to automation within the next two decades[105], and Google's DeepMind has already developed AI that can use its "imagination". According to the company's researchers, the Imagination-Augmented Agents (I2As) uses an "imagination encoder" to make predictions about their environment. When applied to puzzle games, the I2As outperformed regular AIs to a significant degree, but we're a long way away from omnipotent and omniscient AIs coming to subjugate the human race.[106]

But I doubt it will be as bad as that. AI is highly skilled at certain specialized tasks, but we won't see the software taking over jobs that are fundamentally human. Don't just take my word for it – take it from Neil deGrasse Tyson. In a stunning YouTube video in partnership with Mashable, the celebrated astrophysicist weighed in on the debate of whether to send robots or people into space. Tyson believes that it's an irrelevant question. Here's what he said:[107]

Split the question into two parts. Are you only interested in scientific discovery? Send robots. It's cheaper. You don't have to bring them home. If you only care about science then there's no rational reason to send humans, really. For the cost of sending a human, you can send 100 robots. And robots are getting better, smaller, cheaper, faster, smarter, all of the above. But here's the catch. I've never seen a ticker tape parade for a robot. I've never seen a high school named after a robot. I never saw a kid read a book about a robot and say, gee, I wanna be that robot one day. There's value in human inspiration. It's less tangible than the scientific results of an experiment. It's more emotional. It's more philosophical. It's more cultural. It's more human.

---

[105] See: http://bit.ly/automationcalculator

[106] See: http://bit.ly/deepmindai

[107] See: http://bit.ly/robotsorpeople

And humans can deduce things way faster and more intuitively than robots can, at least for now. We can see something we've never seen before and then interpret it. Computers currently only think the way we've programmed them to think. Take a robot and take a human and have them confront something they've never seen before. Whose analysis are you gonna want first? It's gonna be the human analysis, at least for now.

Maybe the day will come when robots will be smarter than humans. But it's very hard to program a robot with curiosity for something it's never seen before. How to process information it knows nothing about. And if you program a robot that has consciousness and curiosity, is it even ethical to send that robot on a space mission with no intent of ever bringing it back? Are you killing a sentient being? So maybe there's a whole frontier we have yet to breach.

At first, that may sound like it has little to do with the world of artificial intelligence in healthcare, but many of the same arguments apply. Picture a future in which healthcare technology has progressed to the point at which human doctors are irrelevant. Would you rather receive a devastating diagnosis from a human or from a robot? And how many people would prefer to be operated on by a human surgeon instead of a robot – even if the robotic operation was statistically less likely to lead to complications?

We're entering a period of history in which a new type of ethics is coming to the forefront. If AI progresses to the point at which it's indistinguishable from a human, should it have the same rights? For the technology to continue to progress, we may need a ratified bill of rights for AI software that's designed to protect friendly AI while shutting down unfriendly AI for the greater good of the world as a whole.[108]

---

[108] See: http://bit.ly/friendlyai

MORE HEALTH IMPLICATIONS OF AI

You could fill a whole library with books about how artificial intelligence is changing the healthcare industry, so keeping it all to a single chapter is a challenge. Here are just a few more ways that AI is being used by healthcare professionals:

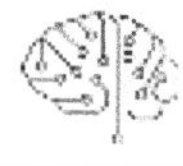

AICURE: Supported by The National Institutes of Health, this smartphone app couples AI technology with the device's camera to confirm that patients are adhering to their prescription. When patients forget to take their medication, or when they don't follow the instructions provided, it can have a huge impact on their outcomes. This app aims to combat that, and it's also being used in clinical trials to ensure the validity of their results.[109]

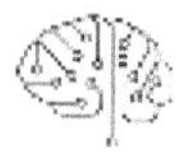

ALS TREATMENT: AI is being used to fast-track the race to treat amyotrophic lateral sclerosis (ALS), also known as motor neurone disease or Lou Gehrig's disease. Richard Mead of the Sheffield Institute of Translational Neuroscience, who's already using AI in his research, said that "many doctors call [it] the worst disease in medicine and the unmet need is huge." Meanwhile, at Arizona's Barrow Neurological Institute, IBM's Watson found five new genes linked to ALS in just a few months, rather than in the "years" it would have taken without it.[110]

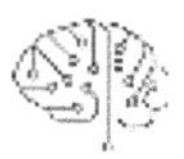

DR. AI: This AI-powered physician translates a person's symptoms into personalized recommendations using deep learning algorithms and advice from a network of over 107,000 doctors. It's powered by HealthTap, a

[109] See: http://bit.ly/aicuresite

[110] See: http://bit.ly/aiforals

World Economic Forum Technology Pioneer, and connects to hundreds of millions of people in 174 countries.[111]

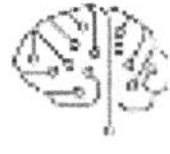

HUMAN AI: The iconic technology magazine WIRED shared the touching story of how James Vlahos preserved the memory of his terminally ill father by creating an artificial intelligence 'Dadbot' that lived on his phone. While this might not seem related to healthcare at first glance, it helped Vlahos to cope with a difficult time and brought comfort to both his terminally ill father and to Vlahos himself. If nothing else, Dadbot was a preventative measure that may have removed or reduced the need for counselling, medication and other aftercare services.[112]

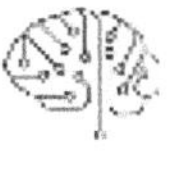

MEDAWARE: This Israel-based organization uses machine learning and AI to find and eliminate prescription errors, which are preventable, costly and dangerous. It works similarly to how financial companies scan for credit card fraud and has just raised $12 million in funding at the time of writing.[113]

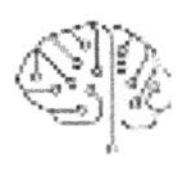

MINDSTRONG: AI startup Mindstrong is led by Dr. Thomas Insel, the former director of the National Institute of Mental Health. It aims to get biometric data (like pulse and activity level) from smartphones to form a picture of users' mental states. If they're angry, they'll treat the device differently than if they were chilled out.[114]

---

111 See: http://bit.ly/drailaunch
112 See: http://bit.ly/jamesvlahos
113 See: http://bit.ly/medawarenews
114 See: http://bit.ly/mindstrongwoebot

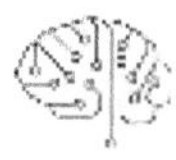

RADIOLOGY ADVICE: Radiologists at the UCLA Medical Center are using an AI chatbot that communicates with referring clinicians and offers evidence-based answers to common questions.[115]

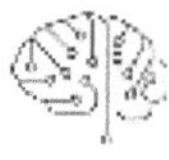

SILVER LOGIC LABS: This AI startup has developed software that uses a camera to identify what people are thinking and feeling. It's being used to provide analytics for TV shows and movies, but it can do much more than that – and all through a typical camera. It can detect whether people are under the influence of alcohol or other intoxicants and it could even identify whether someone's about to have a stroke. CEO Jerimiah Hamon explains, "We did a lot of research and it turns out that if you're going to have a stroke, you'll have a series of micro-strokes first. These are undetectable most of the time, sometimes even to the people having them. If you live at a nursing home for instance, we could have cameras set up, we could detect those."[116]

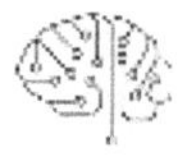

WOEBOT: This AI-backed chatbot helps people to talk about their anxiety and depression, remembering what they say and following up with them over time. Talking about your mood on a daily basis is proven to help fight anxiety and depression, and it also bridges the gaps between visits to a medical practitioner. In one trial, "those in the Woebot group significantly reduced their symptoms of depression over the study period as measured by the PHQ-9 while those in the information control group did not."[117]

---

[115] See: http://bit.ly/uclaaibot

[116] See: http://bit.ly/camerabasedai

[117] See: http://bit.ly/woebotstudy

WILL AI TAKE OUR JOBS?

The short answer is no. Remember what Neil de Grasse Tyson said.

Still, AI does have huge potential to *change* our jobs, and that's where we need to pay attention. Take the use of AI to discover new drugs. So far, there are no AI-inspired, FDA-approved drugs on the market, but the use of artificial intelligence certainly speeds up R&D efforts and helps to predict potential safety and efficacy.[118] It's just that humans are still needed to make intelligent decisions on the data and to carry new drugs through clinical trials before bringing them to market.

People have feared that AI will take human jobs ever since the phrase "artificial intelligence" was first coined in 1956.[119] So far at least, we've seen little evidence to back the claim.

Part of the reason for that fear is that as homo sapiens, the only surviving species of the genus homo, we're afraid that our relative superiority over other living animals will go down the drain if we fork over some of our esteemed cognitive properties to a machine. However, being slightly more intelligent as a consequence of our position on the evolutionary tree is exactly why we *should* work with artificial intelligence.

In 2009, Harvard psychology professor Marc Hauser reported the existence of a large "mental gap" between humans and other animals which he summarized into the four unique aspects of human cognition that are shown below:[120]

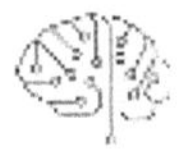

GENERAL COMPUTATION: This is what enables us to create concepts, words, ideas, mathematical formulae and equations.

---

[118] See: http://bit.ly/aidrugs

[119] See: http://bit.ly/aicoined

[120] See: http://bit.ly/themindhauser

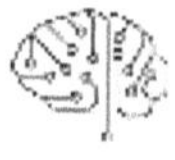

PROMISCUOUS COMBINATIONS: The ability to take a diverse range of topics (such as sex, personal space and friendships) and to use them as the basis of a set of societal norms.

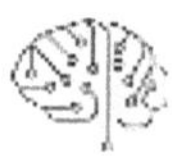

SENSORY COMMUNICATION: Humans use mental symbols to encode sensory experiences (both the real and the imagined) through words, pictures and text. This forms the basics of more complex communication such as language, art, music and software.

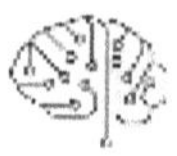

ABSTRACT THOUGHT: This allows us to think about things beyond our current sensory sensations – so above and beyond what we can see, taste, hear, touch and smell. Abstract thought is also what allows us to think about the universe, the meaning of life, infinity and other intangible ideas.

At its most basic level, artificial intelligence can take care of general computation and machine learning can cover promiscuous combinations. Computation in particular is dull and tedious. If we can offload a ton of utterly worthless brain activities to focus on the things that only a human mind can handle, why wouldn't we?

An interesting example of the reality of AI is shown in a piece by Stuart Dredge in The Guardian.[121] Dredge explained that while artificial intelligence is already creating music, it's primarily for use as unique, free-to-use audio for video creators – the YouTube equivalent of muzak.

Ed Newton-Rex, CEO of London-based AI music startup Jukedek, explained: "A couple of years ago, AI wasn't at the stage where it could write a piece of music good enough for anyone. Now it's good enough for some use cases. It doesn't need to be better than Adele or Ed Sheeran. There's no desire for that, and what would that

---

[121] See: http://bit.ly/dredgeai

even mean? Music is so subjective. It's a bit of a false competition: there's no agreed-upon measure of how 'good' a piece of music is. The aim [for AI music] is not 'will this be better than X?' but 'will it be useful for people? Will it help them?'"

It's a little outdated now, but back in 1998, the Organization of Economic Cooperation and Development released a report on technology productivity and job creation.[122] The paper found that technological progression has been accompanied by higher output, higher productivity and higher overall employment – albeit through a process of "creative destruction" in which jobs are being both created and destroyed by progress.

This can be compared to the introduction of the ATM. This allowed banks to automate much of the banking process and to reduce the number of cashiers in every branch. At the same time, this reduction in costs made it cheaper to open up a new branch, so banks responded by increasing the number of locations. This meant that there were still the same number of cashiers but they were spread out across more locations – and every branch also needs a manager, maintenance staff, cleaners, builders and specialist advisors, and so the net effect was more jobs in the industry overall.

AI AND ELECTRONIC HEALTH RECORDS

Chris Bishop, laboratory director at Microsoft Research Cambridge, recently gave a talk in London where he talked about artificial intelligence's role in transforming the healthcare industry.[123] He points out that healthcare systems around the world are under financial pressure – and that those pressures will only get worse as we get older.

As I've talked about elsewhere, Bishop pointed out that while

---

[122] See: http://bit.ly/oecdreporttech

[123] See: http://bit.ly/chrisbishopnext

we're living longer, we're also living with diseases like cancer, dementia and diabetes. This adds extra pressure to the healthcare system, and AI could be just the tool to ease that pressure and make it easier – and cheaper – for patients to receive the help they need.

Through Healthcare NExT, Bishop and his team at Microsoft partnered with the University of Pittsburgh Medical Center – one of the largest healthcare delivery networks in the US – to use artificial intelligence to help with record-keeping.

EHRs are a great idea in principle, but the fact that they're non-interoperable and difficult to use has taken resources away from the patient and forced physicians to become EHR scribes. A study published in the Annals of Internal Medicine revealed that for every hour a physician spends in direct patient care they had to perform two hours of EHR data entry, with the repetitive task often following them home and taking up their evening.[124]

A WORD OF WARNING

As we've seen when it comes to EHRs, AI can help the healthcare industry to manage its huge amounts of data. This includes collecting, storing and normalizing it, as well as processing it to arrive at conclusions.

Machines excel at dealing with data, which is why Google's DeepMind Health project is training itself up by mining medical records to provide better, faster recommendations. IBM's Watson is doing a great job – and sometimes not so great of a job – in numerous settings as well, but it can only get better as time goes on.

It's important to remember that AI works best when it's overseen by humans, when humans and machines are working together for better outcomes. That's because AI doesn't always behave itself.

---

[124] See: http://bit.ly/annalsstudy

A case in point is the story of "Random Darknet Shopper", a Swiss bot created by !Mediengruppe Bitnik as part of an art installation.[125] The bot was given $100 worth of Bitcoin and programmed to randomly purchase items from Agora, an online marketplace on the dark web. Its purchases included Ecstasy, a Hungarian passport and a baseball cap with a built-in camera. Then it got arrested.

Of course, the AI in question was just doing what it was programmed to do, but it should serve as a reminder that we as human beings and developers have the ultimate responsibility when it comes to a robot's actions. It's a gray area that politicians and policymakers are yet to debate.

It reminds me of *RUR*, a 1920 science fiction play by Karel Čapek. The seminal work introduced the word "robot" to the English language while investigating who's ultimately responsible for the actions of a robot – especially one that's capable of thinking for itself. And robots are just AIs with wheels on.

AI can't quite think for itself yet, but it's getting there. In the meantime, it's up to us to decide where the responsibility lies.[126]

THE GREAT EQUALIZER

According to a PwC report called *Bot.me: A revolutionary partnership: How AI is pushing man and machine closer together*[127], "AI has the potential to become the great equalizer. Access to services that were traditionally reserved for a privileged few can be extended to the masses."

AI developer Kaza Razat, who was quoted in the study, added, "As humans, there's a lot we're not good at. When we're making

---

125 See: http://bit.ly/swissrobot

126 See: http://bit.ly/rurkapek

127 See: http://bit.ly/pwcaistudy

machines that are better at certain things than we are, it's still an extension of us. From an evolution standpoint, there are places where we've reached the end of our capacity."

One of the key areas that the PwC poll identified was the field of treating cancer and diseases. Razat explains, "With the enormous amount of DNA data being recorded today, AI could revolutionize personalized healthcare by analyzing that data. Wearables and ingestibles could monitor and correct human behavior to maximize life expectancy and enhance wellbeing. We've already seen AI successfully identify autism in babies with 81% accuracy and skin cancer with 91% accuracy."

REDUCING WASTAGE

There's a huge amount of wastage when it comes to the healthcare industry. Take pharmaceutical companies. It costs billions of dollars to bring a new drug to market, and much of that is because over 95% of compounds don't make it into the hands of the public.[128] What other industry would accept such a high failure rate?

To get to the bottom of the problem, we need to understand the causes. The first reason why drugs fail is purely a matter of safety or efficacy. They might work in a laboratory, but as soon as they're tested on a live patient, all bets are off. Worse still, the drugs can turn out to be toxic as they did in the "Elephant Man" drugs trial. According to the Daily Mirror, "the human guinea pigs were left writhing and screaming in agony as their temperatures soared and their organs began to fail."[129]

The truth is that we can never fully predict how our bodies will react to certain drugs, and part of that is because we don't fully

---

[128] See: http://bit.ly/deeplearninginai

[129] See: http://bit.ly/elephantmantrial

understand how the body works. We don't even understand why humans blush, laugh and kiss.[130]

Drugs work by binding to proteins in the body, and those proteins are what lower your temperature or affect your bodily functions. Most new drug discoveries come about by finding new molecules which can hit known targets, but if you keep on hitting the same targets then you'll keep on having the same effect. Sometimes you need AI to usher in some innovation.

AI also solves the problem of wastage when it comes to professional experience. Let's say it takes us thirty years to fully train a world-class drug scientist – so that including on-the-job experience, they need thirty years of continuous education. By the time that they've learned what they need to know, it's time to retire or retrain. All of that expertise will be wasted.

Thanks to machine learning, artificial intelligence continues to improve itself over time. Instead of retiring after forty years, it'll continue to learn, and all of that learning could be shared between different AI systems. From the data we're starting to gather, we could be building the foundations of a system that will last for hundreds – if not thousands – of years.

TECHNOLOGY FOR PEOPLE

I was recently lucky enough to read global management consulting and professional services firm Accenture's 2017 digital health technology vision report, *Technology for People*.[131] It's a fantastic read whether you're a healthcare professional or not, and I'm not ashamed to say that this section borrows heavily from the report and its findings. Please bear with me here because it's important.

---

130 See: http://bit.ly/unsolvedmysteriesofthebody

131 See: http://bit.ly/dhtvreport

The authors of the report explain, "The unprecedented era of technology innovation is allowing clinicians and service workers to broadly apply their knowledge, freeing up more face time to spend delivering a human touch. And, as technology affords greater opportunities for self-management, it's empowering people to consume healthcare on their own terms. It's no longer about what technology can do for people, it's what people can do with technology."

The report identified five key ways in which the healthcare industry will be disrupted, and I've summarized their findings below:

AI IS THE NEW UI (USER INTERFACE): The increasing sophistication of artificial intelligence means it's able to have a greater impact on our lives, augmenting decision making for clinicians. The report explains, "The relationships between healthcare organizations and people will never go away. But AI will play a primary role in making those relationships stronger through new AI-driven services that help curate, advise and orchestrate lifestyle and care for people."

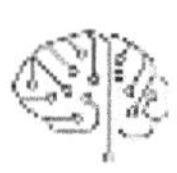

ECOSYSTEM POWER PLAYS: Put simply, healthcare is more open than ever before when it comes to disruption from third-parties, most notably Amazon, Google, Microsoft and other big players. As part of this, the industry is starting to recognize that these tech giants will play a huge role in the future, and so they're mapping out a future in which they can play a part in the transformation of healthcare. According to the report, "90% of health executives believe it is critical to adopt a platform-based business model and engage in ecosystems with digital partners."

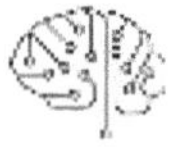

WORKFORCE MARKETPLACE: The way that we work is changing, and more and more people are starting to take on freelance roles or raising money from crowdsourcing. Of course, the healthcare industry is less flexible than others, but it may well be a case of needing to move with the times. The report says, "There are some possibilities for clinical work to be sourced online, but it is more challenging than for back office roles given regulatory barriers from labor laws in healthcare designed to ensure safety. The possibilities for efficiency and productivity are great, but questions remain. The rules must catch up because the workforce marketplace is evolving at speed." I'll investigate the impact of this later in the book when we take a look at disruption from the patients' perspectives.

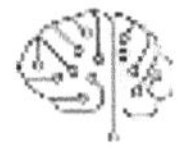

DESIGN FOR HUMANS: Modern technology is designed with humans in mind, but the healthcare industry has been slow to catch up. Once the technology becomes more human-centered, it benefits consumers, clinicians and administrators alike, giving patients the ability to monitor their own health. In the report, Accenture explains, "If someone has a weight loss goal, a health app can suggest foods to eat or offer motivation to exercise. Data can be shared with the person's doctor so that they are informed of health changes. Clinicians can augment the service, as needed, to help encourage the weight loss."

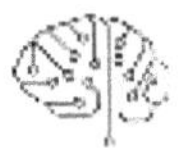

THE UNCHARTED: The biggest question for future physicians will be what the new rules of the game are. Healthcare is a highly regulated industry, and leaders will be needed to fight the good fight to pass new legislation. The report ends on a transformative note: "Healthcare enterprises are not just creating new products and services. They're shaping new digital industries."

CHAPTER SEVEN: MACHINE LEARNING

"MACHINE LEARNING AND ARTIFICIAL INTELLIGENCE MAKES KNOWLEDGE OUT OF A HUGE COLLECTION OF BIG DATA. IT'S LIKE AN ARTIFICIAL BRAIN AT WORK ON THE DATA, INSTEAD OF JUST A STORAGE SYSTEM."

- ERIC XING, PROFESSOR OF THE MACHINE LEARNING DEPARTMENT AT CARNEGIE MELLON

A GREAT DEAL HAS BEEN WRITTEN in recent years about the perils of automation. With predicted mass unemployment, invasion of privacy, declining wages and increasing inequality, we should all be afraid. Right?

Wrong.

Robots aren't coming to steal our jobs – they're coming to take over the stuff that they're better at, like dull, repetitive tasks and massive computations, so that we humans can focus on…well, being human.

Machine learning is the process of teaching computers to teach themselves, and it's an emerging field that's a combination of science and art. It doesn't always work out, like when Microsoft's Tay robot went haywire and learned to be an "evil, Hitler-loving, incestual sex-promoting, 'Bush did 9/11'-proclaiming" robot.[132] But for the most part, machine learning is a useful tool for augmenting the

---

[132] See: http://bit.ly/taygoeshaywire

capabilities of intelligent software, and it has huge implications for the future of healthcare.

Just one example comes to us via a team of interventional radiologists at the University of California at Los Angeles (UCLA), who are developing a machine learning application that guides patients' care. According to an article by Science Daily, "Deep learning was used to understand a wide range of clinical questions and respond appropriately in a conversational manner similar to text messaging. This research will benefit many groups within the hospital setting. Patient care team members get faster, more convenient access to evidence-based information; interventional radiologists spend less time on the phone and more time caring for their patients; and, most importantly, patients have better-informed providers able to deliver high-quality care."[133]

INTRODUCING MACHINE LEARNING

Machine learning is exactly what it sounds like – the ability for a machine to learn, instead of to simply be programmed. Harvard Medical School Assistant Professor and CEO of Cyft, Dr. Leonard D'Avolio, put it best when he said, "It's an objective set of technologies that applied to the right tool can lead to answers to questions that we can't currently answer with our current approaches."[134] In many ways, it's a whole new way of thinking – and one which can't be carried out by the human brain.

Machine learning works by analyzing huge data sets and using them to draw conclusions. Some systems allow user input so that operators can provide feedback to the software, while others rely solely on an algorithm. The algorithms are important of course, but the algorithms can't work without data. It's actually the data that

---

[133] See: http://bit.ly/ucladeeplearning

[134] See: http://bit.ly/bigdatadavolio

allows these systems to be so intelligent.

Historically, machine learning was a slow burner because the academic interest wasn't matched by the commercial potential, but things have changed dramatically over the last twenty years or so. Now we're starting to see academics and commercial interests working together for mutual benefit, and that's no bad thing if it drives the adoption of machine learning technologies.

Machine learning algorithms require two key components if they're to function effectively. The first is a strong model which will enable the software to interpret real-life data in such a way that the algorithm can understand it. The second is a strong pool of data to act as a starting point.

It's actually the models that we're short of, rather than the data. After all, we don't exactly have a shortage of data, and the internet of things is set to increase the amount of data that we gather and store almost exponentially. But the increasing amount of data that we have access to is a double-edged sword. Yes, it allows machine learning algorithms to further improve themselves, but it also requires ever-increasing processing power to be able to operate at full speed, particularly for larger neural networks.

Honestly, our species' ability to capture and store data has always outpaced our ability to process and understand it. Luckily, that's not necessarily a bad thing – after all, the data has to come first, and at least the next generation of machine learning developers will have huge pools of healthcare-related information to train their software with.

REDEFINING CLINICAL TRIALS

To redefine what clinical trials should look like, we first need to consider what they look like at the moment. As a general rule, there are four main types of clinical trial:

PHASE I: Studies which assess the safety of a drug or device.

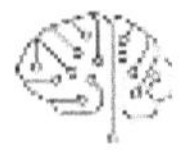

PHASE II: Studies which assess the efficacy of a drug or device.

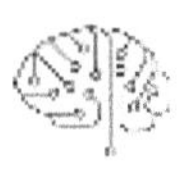

PHASE III: Studies which involve randomized and/or blind testing in several hundred to several thousand patients.

PHASE IV: Also called Post Marketing Surveillance Trials, Phase IV studies are conducted after a drug or device has been approved for consumer sale.

Before a drug gets approved for general usage, it has to successfully undergo a phase III randomized clinical trial. Let's say you've developed a drug to combat heart disease. You'd say, "I need patients between 18 and 65 years of age with a certain blood pressure and specific symptoms." You'd control all of the variables and trial the drug in the perfect world. But the real world isn't perfect – I hate to break it to you.

This reminds me of *Inferior* by Angela Saini, a book which covers the role that women have played in the evolution of modern science.[135] In it, she explains, "Until around 1990, it was common for medical trials to be carried out almost exclusively on men. There were some good reasons for this. 'You don't want to give the experimental drug to a pregnant woman, and you don't want to give the experimental drug to a woman who doesn't know she's pregnant but actually is,' says Arthur Arnold. The terrible legacy of women being given thalidomide for morning sickness in the 1950s proved to scientists how careful they need to be before giving drugs to expectant mothers. Thousands of children were born with disabilities before thalidomide was taken off the market."

Saini continues to explain that it's usually cheaper to study one sex, so most people migrated to the study of males as a large chunk of the female population – those of reproductive age – were

---

[135] See: http://bit.ly/inferiorsaini

considered to be off the market. This focus on males led to the majority of the data being gathered on only half of the population.

"Another problem is that women may respond differently from men to certain drugs," Saini continues. "Medical researchers in the mid-twentieth century often assumed this wasn't a problem. 'There was a notion that women were more like little men. There was a notion that if this treatment works in men, it will work on women,' says Janine Clayton, director of the Office of Research on Women's Health at the NIH in Washington, DC, which funds the vast majority of American health research."

In fact, women are one and a half times more likely to develop an adverse reaction to a drug than men and when the US government looked at the ten prescription drugs withdrawn from the market between 1997-2000, eight of them posed greater health risks to women than to men.

Don't get me wrong, I'm not complaining about clinical trials. I've organized a few myself throughout the years, and they do have real value when they're designed and executed correctly. But they also have their limits, and we're about to head into an era where life itself will be a clinical trial.

Think about it. While there'll still be a need to conduct randomized controlled trials before a drug is released to the public, we'll also be able to track its effects once it's released to the population at large. This is called real world evidence generation. As more and more data is gathered about our day-to-day lives, researchers will be able to monitor the health of people who are taking their drugs and determine the actual long-term impact they have on the population as a whole. And that includes women as well as men. In fact, it includes everyone, regardless of how they identify and regardless of their race, religion or socio-economic background.

Every single patient will be a clinical study. They'll be treated as an individual with personalized medicine, and then their data will be aggregated so we can get an overview of what's really happening when that particular patient takes the medication. We won't just be testing new drugs on a dozen people and blindly prescribing medication – we'll carry out real world evidence trials, which will be more effective and comprehensive, with better outcomes too.

MAKING SENSE OF THE DATA

There are so many new medical publications being published that no physician could ever keep up with them. Even if they spent all of their time reading, without spending any time with their patients, they still wouldn't be able to keep up with it. What chance do they have of staying on top of the literature when they have a busy practice to run?

Meanwhile, it's all too easy to find studies that show low adherence to clinical guidelines.[136] Providers are either too busy to follow guidelines or they're simply unaware of them. Either way, it isn't good news.

Doctors are overwhelmed by the amount of data that they have access to. They're asking how they can take all of that data from the real world and connect it to an electronic health records system. Meanwhile, every single day, there are thousands of new studies coming out and the average human just doesn't have the time to keep up with them all.

On top of that, every patient is different. Everything they do in their lives – where they live, whether they smoke and what they eat – builds up over time to create an overall disease. There are factors that we don't even know are factors. Let's assume that eating breakfast at 7:30 AM reduces your chances of bowel cancer when compared to eating it at 7:45. We just wouldn't know it.

That's where machine learning comes in. Computers can do things that humans can't, and one of their main advantages is that they can process calculations at a phenomenal rate. A human being might not be able to make sense of the data, but a machine could – as long as it was well-programmed and designed to use machine learning techniques to teach itself along the way.

Rebecca Herson, Vice President of Marketing for real-time analytics company *Anodot*, explains, "There's just so much data

---

[136] See: http://bit.ly/clinicalguidestat

being generated, there's no way for a human to go through it all. Sometimes, when we analyze historical data for businesses we're introducing *Anodot* to, they discern things they never knew were happening." The company has repeatedly found that 80% of the anomalies identified by its machine learning software are negative factors as opposed to positive opportunities. They were losing money due to inefficiencies, not missing out on potential revenue.[137]

None of this technology is new. Personalized medicine follows the same concept as Facebook, Netflix and other popular websites. Netflix knows what you like and what you don't like, and it makes recommendations based upon that. Medical software would work in a similar way, except it would tap into medical data and you'd have a trained physician overseeing it. Humans and machines partnering for better outcomes.

The crazy thing is that healthcare providers are already using Facebook, Netflix and Amazon on a personal level, but many are still afraid of machine learning when it comes to healthcare. They think, "Oh no, the computer's coming, that means I'll be out of a job." What we're actually talking about is having enough data to make the best decision possible for the patient. You're not spending time asking questions and operating on guesswork – you're making informed decisions based on aggregated data and the personal healthcare profile of every patient.

People just don't realize that companies like Netflix and Facebook are using the same technologies that could revolutionize healthcare if applied to it.

When we talk about companies like Fitbit and Apple with the Apple Watch, it's not the devices that are important – it's the data. Imagine if Apple started to give out free smartwatches to capture data. If that happened, they'd spawn an entire ecosystem around them because there'd be so much data and so many potential applications. Apple would have free access to everyone's health data, creating a monopoly and automatically turning them into the

---

[137] See: http://bit.ly/anodotai

largest healthcare company in the world.

Of course, Apple won't do that (although it's not like they don't have the resources). They're not focusing on healthcare, they're focusing on making a profit, which is why they're charging $500 per device. But they're also focusing on the data that these devices can capture, and they know that by gathering it now, they set themselves up to play a pivotal role in the healthcare industry of the future.

It's also interesting to note Apple's commitment to using machine learning to make sense of data. The company has even launched a blog with the catchy title of the "Apple Machine Learning Journal"[138], ostensibly to break their usual veil of secrecy and to remind the world that they have the tools and resources to become a key player in the world of big data and machine learning.

You only need to take a look at Facebook and Google to see how there's a huge amount of potential for companies to monetize the data they hold. The problem is that investors want to make money, and the data-based approach can seem dangerous and intangible – but only if you don't understand the business model.

Perhaps Jenna Wiens, whose work focuses on reducing preventable deaths from hospital-acquired infections like C. difficile, puts it best. After being named as one of MIT's 35 Innovators Under 35, she told the Technology Review: "I think there is a bigger cost to not using the data. Patients are dying when they seek medical care and they acquire one of these infections. If we can prevent those, the savings are priceless."[139]

OVERCOMING HUMAN BIAS

Machine learning systems are deliberately designed to overcome human bias, but that doesn't always mean that they're successful.

---

[138] See: http://bit.ly/amljournal

[139] See: http://bit.ly/jennawiens

When we talk about human bias, we talk about the natural tendency that people have to subconsciously add their own beliefs and opinions to a piece of software. So if an Australian programmer creates a voice recognition system, it's likely to work better with Australian accents than with French accents.

Machine learning aims to overcome this by allowing machines to 'teach' themselves – so in this case, if the Australian's software used machine learning then it would get better and better at understanding French accents the more it 'spoke' to French people.

But that still doesn't allow us to fully overcome the bias. There's a great example in Google's *Machine Learning and Human Bias*[140] video. When people are asked to draw a shoe, they're much more likely to draw a Converse-style shoe than a pair of high heels, which led to the software recognizing one type of shoe but not another.

You might think that machine learning technology can't be exposed to bias because it's designed to find patterns in data. Unfortunately, just because something is based on data, it doesn't mean that it's neutral. Even when we start out with the best of intentions, our unconscious prejudices naturally creep in.

Imagine using machine learning to build a profile of a typical physicist. If you were using past data, it would be inherently biased towards men because of the historical patriarchy, but that would be of little to no use when predicting trends for the future.

This presents an interesting challenge for programmers because machine learning is becoming such a vital part of our lives, and not just when it comes to the future of healthcare. Developers will need to make a concentrated effort to prevent new technologies from inheriting our human biases.

---

[140] See: http://bit.ly/machinelearninggoog

THE PROBLEM WITH WATSON

We've talked a lot about IBM's Watson throughout the book, but not everyone is convinced that it lives up to its impressive reputation. I recently met up with Scott Parfitt, a psychophysicist and entrepreneur and the CEO of Content Technologies, Inc. Parfitt is an AI and deep learning genius (no exaggeration) who's based in Los Angeles, and he thinks that Watson is "overhyped ".

When we first met up, he told me that his system is far superior to Watson, and I've come to agree with him over time. The issue with applying Watson to healthcare is that it's trying to solve the wrong problem. The name of the game is outcomes. No, more than that – the name of the game is value, which comes from a mixture of both patient outcomes and the cost it requires to deliver them.

Parfitt used a great analogy to describe the state of medical intelligence, explaining: "The parable of the blind men and the elephant originated in the ancient Indian subcontinent, from where it has been widely diffused. It's the story of a group of blind men, who have never come across an elephant before and who learn and conceptualize what the elephant is like by touching it. Each blind man feels a different part of the elephant's body, but only one part, such as the side or the tusk. They then describe the elephant based on their partial experience and their descriptions are in complete disagreement. In some versions, they come to suspect that the other people are dishonest and they come to blows. The moral of the parable is that humans have a tendency to project their partial experiences as the whole truth, and that one should consider that one may be partially right and may only have partial information."

The problem with Watson is that it's exactly like the blind men trying to understand the elephant, and it's not just Parfitt and myself who are disappointed by Watson's failure to fully deliver. Casey Ross and Ike Swetlitz of Stat penned a damning article in which they investigated the impact of Watson on cancer care, ultimately determining that it's "nowhere close" to the original vision of "a revolution."[141] There are several reasons for this, from a lack of adoption to complaints that its advice is biased towards American patients and methods of care.

Ross and Swetlitz explain, "The interviews suggest that IBM, in its rush to bolster flagging revenue, unleashed a product without fully assessing the challenges of deploying it in hospitals globally. While it has emphatically marketed Watson for cancer care, IBM hasn't published any scientific papers demonstrating how the technology affects physicians and patients. As a result, its flaws are getting exposed on the front lines of care by doctors and researchers who say that the system, while promising in some respects, remains undeveloped."

IBM claimed that Watson for Oncology could use artificial intelligence to analyze data and to generate new insights and "even new approaches to cancer care." But STAT found that "the system doesn't create new knowledge and is artificially intelligent only in the most rudimentary sense of the term." They conclude that "its programming is akin to a different game-playing machine: the Mechanical Turk, a chess-playing robot of the 1700s, which dazzled audiences but hid a secret – a human operator shielded inside."

Despite this, healthcare remains an important part of IBM's strategy. They employ 7,000 people in their Watson health division and believe the industry will grow to $200 billion over the next few years. That's a pretty big gamble for the company to make.

In an article for CNBC, Christina Farr – who specializes in covering "the tech/health convergence" – explained the situation by saying, "IBM Watson might not deliver on its lofty goal to revolutionize medicine with artificial intelligence, as a recent investigation suggests, but that doesn't mean new technology will be incapable of improving healthcare."[142]

The problem is that new technologies too often ignore real-world

---

[141] See: http://bit.ly/ibmwatsonarticle

[142] See: http://bit.ly/cnbcwatson

circumstances. Farr says "throwing sexy tech at the problem won't prove useful in the day-to-day operations of a hospital, and it certainly won't replace doctors, as some have claimed."

Ultimately, while the future of healthcare will definitely involve technological innovation, that's not what it's all about. New technologies are only useful if they offer actual value to either the patient or the physician. Otherwise, new tech is just like EHRs – a good idea, but a huge waste of time in practice.

The good news for Watson is that while it has its problems, it also has its advantages. In an article for the MIT Technology Review, David H. Freedman explained, "The best bet for getting the data lies in close partnerships with large healthcare organizations that tend to be technologically conservative. And one thing IBM still does very well in comparison to startups, or even giant rivals like Apple and Google, is gain the trust of executives and IT managers at big organizations. IBM has a crucial advantage. It's getting Watson inside a wide range of medical centers, healthcare administration groups and life science companies, all of which are positioned to provide the critical data needed to shape AI's future in medicine."[143]

Freedman explains that machine learning systems like Watson are trained by "constantly rejigging [their] internal processing routines in order to produce the highest possible percentage of correct answers." But to do this, you need to be able to tell the software when it's right and when it's wrong and so the correct answers need to be known in advance. Freedman says that "the more training problems the system can chew through, the better its hit rate gets."

That's where IBM's access to data will come in handy, but for Watson to be truly effective then it needs access to all sorts of ancillary data that the healthcare industry has been slow to gather, such as whether patients are drug-free, what their diets are like and whether they're breathing clean air or living in a smog-filled city. All of these factors influence the health of patients and, as a

[143] See: http://bit.ly/ibmwatsonpaultang

consequence, the treatments that they require.

Manish Kohli, a physician and healthcare informatics expert at the Cleveland Clinic, put it best when he said, "Healthcare has been an embarrassingly late adopter of technology."

Luckily for us, it's not too late to change.

DEEPMIND AT THE NHS

The NHS might be a role model in many ways, but it's not without its problems. One of the biggest challenges for the NHS is the lack of resources, which is where AI and machine learning come in. Instead of bringing in more doctors, there's the potential to outsource many of their most time consuming tasks to artificial intelligence.

The full power of AI comes not from its ability to process information but from the insights it can learn along the way. That's why the NHS partnered with Google's DeepMind to carry out a number of trials in the controlled environment of British hospitals.

Channel 4 News caught up with DeepMind's founders, who sold the company to Google for $400 million[144] after four years of existence, and the subsequent interview shines a light on the issues that the industry will face.[145]

"What I'm really worried about is that the fear and the reactionary paranoia is going to limit the access to what is clearly going to be an incredibly valuable technology which will change people's lives," explains Mustafa Suleyman, co-founder of DeepMind. The fear isn't exactly ungrounded – a couple of months after the interview, the UK's Information Commission (ICO) ruled that a British hospital "didn't do enough to protect the privacy of

---

[144] See: http://bit.ly/googledeepmind400

[145] See: http://bit.ly/channel4ai

patients when it shared data with Google."[146] The issue was that the Royal Free NHS Foundation Trust "did not tell patients enough about the way their data was used". Although they weren't fined for the discrepancy, the trust said it would tackle "shortcomings" in its data-handling in the future.

But this partnership is actually offering a huge amount of value to patients. The Royal Free NHS Foundation Trust was using 1.6 million patient records to "develop and refine an alert, diagnosis and detection system that can spot when patients are at risk of developing acute kidney injury (AKI)." And kidney treatments are just the start.

Take blindness. In the UK, blindness costs £28 billion every year, equivalent to a fifth of the overall health budget. A major cause of this is macular degeneration, which is treatable when caught early enough. Machine learning is already being used in the fight against macular degeneration thanks to London's Moorfields Eye Hospital.[147] Patients are scanned in what looks like a normal scanner, and its internal machine learning tool can analyze thousands of complex images almost instantaneously, examining the back of the eye and a cross-section of the retina at a higher magnification than you'd get from an MRI machine.

This task would usually take doctors hours on end – but the machine can do it in a couple of seconds.

The implications of this are huge. In the UK alone, 200 people every day are diagnosed with macular degeneration. The NHS aims to identify and treat these people within two weeks, but that target just isn't being met. Opening up machine learning to the rest of the country could allow the NHS to identify those patients and to treat them within a couple of days.

This isn't necessarily lifesaving, but it could dramatically improve the quality of life for hundreds of thousands of people across the country. Meanwhile, doctors would be able to reclaim

---

[146] See: http://bit.ly/deepmindico

[147] See: http://bit.ly/moorfieldseyehospital

their day as less time would be taken up making diagnoses. This is another great example of how humans and machines can partner for better outcomes. Everybody wins, including the patient.

THE SECRET RULES OF MODERN LIVING

In his fantastic documentary, *The Secret Rules of Modern Living: Algorithms*[148], professor Marcus du Sautoy talks about how algorithms and machine learning are already revolutionizing our lives. He explains, "Algorithms are being used all over the world. In Denmark, to match children to daycare places. In Hungary, to match students to schools. In New York, to allocate rabbis to synagogues. And in China, Germany and Spain, to match students to universities."

du Sautoy also talked about how algorithms can impact healthcare by pairing donors with recipients, sharing the stories of two patients who needed transplants and whose loved ones were willing to donate. Unfortunately, their tissue types were incompatible. That's where the NHS came in with its powerful algorithm, essentially matching up pairs of recipients and willing donors to save lives out in the real world.

"When we first looked at this problem, we really underestimated the complexity," explains Rachel Johnson, a statistician for the NHS. "Originally, we just started with swaps between two pairs, so it was very simple. But it soon became obvious that we needed something much more complex. There are 200 patients in each of our matching runs. We need to look for all of the possible transplants and it's surprising how many there are. There are hundreds – sometimes thousands – of possible matches. It's just something that wouldn't be possible without the algorithm."

This is where machine learning and artificial intelligence comes

---

[148] See: http://bit.ly/secretrulesalgorithms

in. It's able to process huge amounts of data that a human being just couldn't hope to get through, and it gives surgeons practical advice that can save lives without taking jobs away from them.

The NHS program has already helped to give a new lease of life to 400 patients, despite still being in its early days. One recipient told filmmakers, "Many years ago I wouldn't have had this chance. I feel very grateful to [the donor] – and also to the algorithm."

THE ETHICS OF MACHINE LEARNING

Ultimately, it comes down to this. Artificial intelligence and machine learning allows us to capture the training and experience of thousands of people very quickly and to apply that experience to the huge amounts of data that we're accumulating on a daily basis.

When it comes to fetus scans, we capture more imagery than any single person could ever look through in a lifetime. Our systems are overwhelmed because millions of pictures are being created at every single center in the country. Human beings could never process all of the imagery, but AI can – and it could help to identify any early warning signs that a human physician can then further investigate.

This isn't a case of artificial intelligence stealing human jobs. After all, humans aren't looking at all of those images in the first place, so AI is simply a tool that takes on a job that we have neither the time nor the inclination to handle.

We've already talked about whether humans will ever be able to accept mistakes that are made by machines, but when it comes to healthcare, it may just be something that we need to figure out as a society.

Just take a look at the Stanford researchers who developed an algorithm that can diagnose melanomas "as well as or better than expert dermatologists."[149] That might sound like a good thing, but it

---

[149] See: http://bit.ly/medicalliability

still has an accuracy rate of less than 75%, so it's easy to understand the moral dilemma. If a human doctor were to take the patient on, they'd also accept responsibility. If AI software was used, who'd take responsibility in those rare cases where it got it wrong?

I like to think that there's a middle ground, where humans and machines can partner for the best possible outcomes. Two pairs of eyes are better than one, and new technologies could offer a built-in second opinion in which the machine checks on the doctor and the doctor checks on the machine. If there's a discrepancy between the two, a third opinion could be sought or new tests could be run. In the end, humans and machines will only get better and smarter.

It's not far-fetched to say that in the coming years, it may be negligent for a doctor to refuse the help of AI because it could have a damaging effect on their patients' overall prognosis. In the future of healthcare, artificial intelligence might not be optional.

CHAPTER EIGHT: THE FUTURE FOR PATIENTS

"ALL OF OUR ONLINE INTERACTIONS TELL US SOMETHING IMPORTANT ABOUT OUR HEALTH."

- JOHN BROWNSTEIN, HARVARD MEDICAL SCHOOL PROFESSOR AND RESEARCHER

NEW TECHNOLOGIES can change the way the world works. Just look at fire. The wheel. The telephone.

But new technologies don't just bring advantages. Social networking in particular has changed the way we look at healthcare thanks to its often harmful effects on mental health. In his 2016 book *Social Paranoia: How Consumers and Brands Can Stay Safe in a Connected World*[150], author (and my editor) Dane Cobain cited a number of psychological studies that showed a connection between social networking use and common disorders such as anxiety and depression. A 2017 study identified Instagram in particular as having a negative impact on young people's mental health.[151] Meanwhile, it's becoming increasingly common for online celebrities to come out as sufferers or supporters of mental health issues.

One such celebrity is Taryn Southern, one of the early YouTube stars. By the way, if you think that a YouTube star can't be a

---

150 See: http://bit.ly/socialparanoia

151 See: http://bit.ly/instagramformh

celebrity, guess again. YouTube users watch over a billion hours of video every day, and it's set to overtake US TV viewership within the next few years.[152] Southern realized that the pressure to continue creating content could have a serious impact on her way of life, as well as her mental health. "The only way to feed the beast is posting daily or creating multiple channels for niche formats," she said. "You just feel like a slave to the algorithm."

Southern eventually resorted to cutting back on uploads and spending more time offline. She started to see a therapist and discussed her mental health with her online following, echoing the trend of speaking freely about mental health issues – a trend which I hope will continue. When people talk about healthcare – whether mental or physical – they need to take ownership. Talking freely about mental health issues is just the start when it comes to ushering in the future of healthcare.

RETHINKING HOSPITALITY

For the patient, the disruption of healthcare will make the entire experience less unpleasant, whether it's by providing personalized healthcare that stops you from being ill or whether it's by creating an infrastructure in which the hospital is designed from the ground up to help the patient.

Bert Greenstein, vice president of IBM's Watson, says that medical staff spend nearly 10% of their time answering questions about lunch, visiting hours and other hospital policies.[153] Imagine if those questions could be answered by a voice assistant like Siri or Alexa – and what the nurses could be doing with their time instead. Imagine if bedbound patients could ask Siri to open the curtains, or if Alexa could monitor the patient and keep the room at their ideal

---

[152] See: http://bit.ly/youtubetakesover

[153] See: http://bit.ly/watsonhospital

temperature.

This may sound far-fetched, but Alexa is no stranger to hospitality and healthcare. It also often shows a surprising understanding of the world around it, such as when it was credited with making a potentially lifesaving 911 call during a domestic dispute.[154]

AI bots and voice assistants are often thought of as being better suited for younger users, the digital natives who don't remember a time without technology. But the truth is very different, and systems like Alexa are increasingly being used by older users – and by women in particular. One study by Verto found that the average "superuser" of personal assistant apps is a 52-year-old woman who spends 1.5 hours a month using the technology.[155]

This is a good sign for the use of AI systems to provide better hospitality, whether you're checking into a hotel or whether you're visiting the hospital. The technology is still in its infancy, but it'll be interesting to watch it and to see how it develops.

A FRIENDLY EAR

Voice assistants won't just change the way we interact with our devices. They'll make their way into grocery stores to help shoppers, hospitals to help patients and their visitors and nursing homes to help the elderly.

But for them to achieve their full potential, they'll need to become better conversationalists. At the moment, the best they can manage is to tell a joke on command or to respond in humorous ways to a set of pre-programmed questions, but if machine learning technology can be deployed en masse to turn these assistants into better conversationalists, their benefits will become even more

---

[154] See: http://bit.ly/alexa911

[155] See: http://bit.ly/averagevoiceuser

apparent.

If you're a 90-year-old widower who lives alone then you'd get a huge amount of value from Alexa if she was able to chat to you about the weather. You'd also be able to ask your assistant what family members are up to and to get a response from their social networking feeds, and you could even reminisce about the past if it had access to historical data.

A recent survey found that three quarters of older people in the UK are lonely[156], and both loneliness and social isolation are important health risks when it comes to the elderly.[157] Loneliness can be devastating, and virtual assistants could be our answer to the problem. They could also come to the aid of mental health patients, which is particularly important when you consider that suicide is the biggest single killer of men aged under 45 in the UK.[158] Any tool that can offer support and improve outcomes will be invaluable in the fight for better healthcare.

Better yet, these voice assistants will help patients to book appointments, to email specialists and to read out their responses. They'll be able to continuously monitor vulnerable patients and remind them to take their medication. They'll also keep an eye on lifestyle factors and offer up personalized advice to patients to help them live a healthier lifestyle.

I'm deeply interested in the study of loneliness and depression in the elderly, as well as chronic disease and the impact it has on their health. After this book has been published, I'll be working on a social media platform (LIVYANA) that's designed to connect patients with other, similar patients so they can chat with each other. Diabetes sufferers could talk to other diabetes sufferers. Elderly patients could do the same, and so could mental health sufferers or cancer patients. Their carers could also get involved and share tips on how to best beat the illness.

---

[156] See: http://bit.ly/lonelyuk

[157] See: http://bit.ly/lonelinesselderly

[158] See: http://bit.ly/maledepressionstat

I'm convinced that by creating a virtual support network for people with similar conditions, we can help with patient education, loneliness and social isolation while improving patient outcomes across the board.

VIRTUAL ASSISTANTS IMPROVING PATIENT CARE

You might think that voice assistants in physical care are a far-flung thing of the future, but we're already starting to see their uses in clinical applications.

Take Molly. Molly is the world's first virtual nurse, developed by medical start-up Sense.ly to help people to monitor their condition and treatment plan.[159] It uses machine learning to support patients with chronic conditions between visits to the doctor.

Another great example comes to us from Boston Children's Hospital, who launched an Amazon Alexa app called KidsMD. The idea is to allow Alexa to offer health advice to parents about children's illnesses and medication. An article from Mobi Health News explains, "Any Alexa-enabled device, including the Amazon Echo, Echo Dot, Amazon Tap and Amazon Fire TV, can download the app. Once it's installed, parents will be able to ask Alexa whether symptoms like fever, cough, headache, rash, vomiting, sore throat, diarrhea, fatigue or shortness of breath warrant a call to the doctor. They can also ask about weight or age specific dosing guidelines for over-the-counter drugs like acetaminophen."[160]

Boston Children's Hospital also has an innovation center which brought fifty people together to brainstorm new ways that Amazon's Alexa could help in the clinic and at home. The idea was to "[work] to bring Alexa into patient rooms, help doctors take notes and read back charts, among other things," and the technology was

---

159 See: http://bit.ly/virtualmolly

160 See: http://bit.ly/kidsmdstory

demonstrated on electronic equipment in an operating room, an intensive care unit and a child's bedroom."

"In the ICU," one article explains, "visitors saw how Alexa could help nurses. Each time they draw blood, they have to figure out how much to draw and choose which of many color-coded vials to put it in. Nurse Paula Lamagna estimated she could save 15 to 30 minutes per patient if she used a voice-activated system to get that information."[161]

Paradoxically, a reduction in the amount of time it takes to prepare for each patient could actually lead to nurses spending more time with them.

GREATER COMFORT

Illnesses are unpleasant. It can be a stressful time, especially if a patient is suffering from a terminal disease or facing long-term debilitation. That's why the future of healthcare will see a greater emphasis on the comfort of patients – and on emotional comfort, as well as just the physical.

Emily is a Facebook Messenger bot that was developed by LifeFolder.[162] It's designed to talk to people about death and end-of-life decisions, such as who you want to speak for you if you're unconscious and whether you want to be an organ donor. A conversation with Emily takes around half an hour and will leave you with legally-binding documents, making it a pain-free process to give patients peace of mind. Let's say you've been diagnosed with terminal cancer and given three months to live. A quick chat with Emily would help you to make some sense of the situation that you're in, even if that's only by helping to save lives with your organs once your time on earth is over. The most important benefit of a service like Emily is also the most intangible. Put yourself in the

---

[161] See: http://bit.ly/alexaatboston

[162] See: http://bit.ly/emilybot

shoes of that terminal patient. Wouldn't you rather get the admin out of the way so you can spend the rest of your life with friends and family?

While artificial intelligence and voice/virtual assistants could offer a huge amount of comfort to patients, medical facilities face the added complication of legislation, which could derail these initiatives before they even get started. That said, they're already being tested by IBM in partnership with Pennsylvania's Jefferson Hospital, where patients may soon be able to "convalesce with greater comfort and convenience, directing their care teams by ambient voice command."[163] The idea is to create smart hospital rooms that are powered by the internet of things, making it easier than ever for patients to request information or to modify their environment by dimming lights, adjusting temperatures or turning on music.

"Being in a hospital can often be a hectic, anxiety-ridden, or even an intimidating experience for patients and their loved ones," explains Neil Gomes, vice president for technology innovation and consumer experience at Thomas Jefferson University and Jefferson Health. "If we can minimize that discomfort, even a little, we are doing a lot to increase the well-being and care of our patients."

This kind of technology is already in use in hotels all over the world. In the leisure industry, internet-connected technology is there to make the visitor's stay as comfortable as possible – and when hospitals switch to a value-based model, they'll put a similar effort in to make their patients' stay as comfortable as possible.

One such innovator is Yotel, which introduced the world's first robot concierge. ZDNet explains, "[They] collaborated with MFG Automation to adapt a robotic arm normally used in factory assembly. The huge machine is behind a glass wall where guests can watch it work."[164] It can even send your luggage directly from the hotel to the airport.

---

[163] See: http://bit.ly/jeffersonhospital

[164] See: http://bit.ly/yobotrobot

Meanwhile, when they check into the Hotel Irvine in California, "Guests can connect many of their gadgets, including laptops, smartphones and tablets, to their room's 42-inch flat-screen TV via the property's new 'myAway' interactive in-room service. The platform allows users to stream favorite TV shows and movies, check their flight status, play personal playlists and check in and out of their room."

"Today's travelers expect up-to-the-minute, customized technology – just like they have at home," explains Jeroen Quint, general manager of Hotel Irvine. "Our new in-room entertainment is versatile, interactive and user-friendly – furthering our quest to customize and enhance the guest experience."[165]

If hospitals are to offer the best possible service to their patients – who are, after all, customers – then they'll need to borrow from the hospitality industry. Besides, if they're comfortable during their stay at a facility then it's going to help them to get better. One study into the impact of physical environmental factors during the healing process found that "health is a state of complete physical, mental, and social well-being and not merely the absence of disease or infirmity."[166]

In the future, hospitals will be hospitable and patient centric. They'll have to be. Our current system is incentivized to optimize volume rather than quality. At the moment, if we want to order takeout or book a holiday, we can look up reviews and ratings on Yelp and TripAdvisor. Won't it be amazing when you can choose which hospital you go to based upon the quality of their care – as determined by other patients, and not by some faceless governing body?

There's growing evidence to suggest that such a shift is already happening, with patients switching doctors based on the services they offer and the quality of care received as opposed to just the costs. In fact, only 32% of patients are completely satisfied with their

---

[165] See: http://bit.ly/cuttingedgehoteltech

[166] See: http://bit.ly/healingenvironmentstudy

provider, while 12% have already left their doctor's practice and 34% are considering leaving.[167] Millennials in particular are more likely to switch doctors, with dissatisfaction caused by everything from inconvenient locations and outdated booking systems to the lack of SMS alerts and email reminders.And so it's clear that patients are expecting more and more from the facilities they use. At the same time, existing healthcare providers are often struggling to meet those demands, causing a growing disparity between what people expect and what they'll actually receive when they engage with a healthcare practitioner.

MEET THE HOSPITAL OF THE FUTURE

It's time to take a trip to Omaha, Nebraska. The city's new $323 million Fred & Pamela Buffett Cancer Center is designed to spur collaboration between doctors and researchers, but it also took the innovative – and obvious! – step of asking patients what they wanted. When patients arrive, they'll find furnishings "tested and selected by other cancer patients, from the heated infusion chairs to the memory foam stretcher mattresses intended to keep patients comfortable during treatment sessions that can stretch up to 10 hours."

The facility has been built with two key objectives in mind: ensuring patient comfort and bringing doctors from different disciplines together to facilitate serendipity, communication and, ultimately, breakthroughs in the way that we understand (and treat) cancer. It was opened by former Vice President Joe Biden, whose son Beau died of brain cancer in 2015 at the age of 46. Biden has said he'll spend the rest of his life fighting to eradicate the disease.[168]

In an article about the facility, Live Well Nebraska explains, "The

---

[167] See: http://bit.ly/patientservicestat

[168] See: http://bit.ly/joebidenatbuffett

integration begins at the front entrance, which will be used by patients, families, clinicians and researchers. It leads into an expansive, art-filled lobby which previews the facility's extensive Healing Arts Program and has a concierge desk. Patients will be able to make one call and get one appointment to see all of their specialists. If they need lab tests, they can visit the adjacent clinical services area beforehand or take an escalator to radiology for an X-ray, eliminating the need to trek from building to building for services. Results will be ready for their appointment. Also inside are a grab-and-go cafeteria and a 70-seat dining room. A separate area with a glass-fronted fireplace and sound-dampening wood veneer panels provides a quiet place for discharged patients to wait for rides."[169]

I've visited the place and it's remarkable. In particular, I was stunned by the Chihuly Sanctuary and the installations by world-renowned glass artist Dale Chihuly. The reflection room in particular is incredible and emotional, and the art installations more than meet the goal of "[providing] a place of respite and reflection for patients, families and staff dealing with cancer."[170]

QUALITY OF LIFE

We're entering a new stage of human evolution. Look at the people who are setting records for being the oldest person alive. Every time they're asked how they lived so long, they have no idea. But if everyone was wearing a healthcare tracker, we could analyze the data from everyone who lived into their triple digits and start to identify what made them live so long. Then we could roll that out across the world and improve global lifespans.

But the future of medicine is more than that. We're not talking

---

[169] See: http://bit.ly/livewellnebraska

[170] See: http://bit.ly/chihulysanctuary

about living longer, we're talking about living better. It you look at the people who live to a hundred and something, they're usually relatively healthy. They might be old, but they're not unwell – and as often as not, they're taking little-to-no medication.

In the future, there's no reason why each and every one of us can't do the same. If you monitor your health along the way and take a preventative approach, you won't have heart disease in your retirement because you took steps to avoid it before it became a problem.

Imagine a thirty-year-old male in America who's on statins for cholesterol. If this male lives to the age of ninety, imagine how much money will be spent on statins and on the associated costs for the patient's healthcare. Now imagine if we could use technology to better manage the patient, such as by offering personalized coaching through Alexa, daily activity tracking through wearables, etc. The patient could live a much longer, healthier, happier life, free from medication and their associated costs and side-effects.

By the way, if you're curious about just how much medication we're taking, check out the photos of Pharmacopoeia's *Cradle to Grave* installation at the British Museum. The 14-meter long exhibit revealed all of the medicines prescribed to one woman and one man during their lifetime in the UK, then laid them out in the exact sequence in which they would be taken.[171]

This out-of-control consumption of medication is unsustainable and it simply can't continue. So here's a prediction. The pharmaceutical industry is going to change its model from selling drugs by volume to selling drug plus support services to the right patient for a premium. Drugs won't be used as a preventative measure for lifestyle-induced illnesses for the young but will be reserved for those who develop diseases in their later years.

Quality of life is a natural side effect of value-based and preventative healthcare. We need to focus on stopping people from getting sick, rather than waiting for them to get sick and then

---

[171] See: http://bit.ly/pharmacopoeiaexhibit

pumping them full of medication.

MENTAL HEALTH AND DEPRESSION

Mental health problems are one of the main causes of the overall disease burden across the world. Major depression is also the second leading cause of disability, causing over 40 million years of downtime in 20 to 29 year olds. In the past week, one in six people experienced a common mental health problem.[172]

On top of that, many of our existing treatment options don't really work. Even the most optimistic analyses of the most effective drugs find that they decrease the probability of depression by only 20%. It turns out that simply moving from Chicago to Honolulu could be twice as effective as medication thanks to the increase in warm weather.[173] This makes sense, because there's evidence that links depression to a deficiency of vitamin D, the sunshine vitamin, with one study suggesting that "effective detection and treatment of inadequate vitamin D levels in persons with depression and other mental disorders may be an easy and cost-effective therapy which could improve patients' long-term health outcomes as well as their quality of life."[174]

It turns out that you there's a huge amount of latent health data buried beneath Google's search history. By analyzing anonymous, aggregate data from tens of millions of search queries, Seth Stephens-Davidowitz, the author of *Everybody Lies: Big Data, New Data, and What the Internet Can Tell Us About Who We Really Are*[175], was able to identify trends around depression related searches that give us an insight into the psyche of the nation as a whole.

---

[172] See: http://bit.ly/mentalhealthstatsworldwide

[173] See: http://bit.ly/drgooglewill

[174] See: http://bit.ly/vitamindanddepression

[175] See: http://bit.ly/everybodyliesbook

Stephens-Davidowitz says, "According to the data, depression is highest in April. T. S. Eliot was probably right to call it 'the cruelest month.' Depression is lowest in August. The state with the highest rate of depression is North Dakota. The one with the lowest is Virginia. The city with the highest rate is Presque Isle, Me. The city with the lowest is San Francisco. The date on which depression is lowest is December 25th, followed by the days surrounding it. The Great Recession appears to have caused a significant increase in depression. In 2009 and 2010, there was a large increase in depression queries in states with large increases in unemployment, like Nevada, Florida and Alabama, compared with states that were little affected, like North Dakota and Arkansas. A 1% increase in the unemployment rate was associated with a 2% increase in depression queries."

He doesn't stop there. He continues to explain, "The strongest predictor by far [is] an area's average temperature in January. Colder places have higher rates of depression, with the correlation concentrated in the colder months. Besides temperature, the biggest predictors of 'depression' searches were the sizes of the Hispanic population and the college-educated population. More Hispanic-Americans meant fewer searches (though this might have been a result of language factors). More college-educated people meant less depression. This finding challenges the stereotype of the overeducated, overanalyzing depressive. There's a lesson here for public health and medical researchers. Are you investigating how weather affects migraine headaches? How chemicals in water affect autism rates? I believe we are about to enter a golden age of disease research. Many of the biggest developments will come from the analysis of big data, not from traditional experiments that survey a relatively small number of people."

This is a perfect example of humans and machines partnering for better outcomes. How else would we have known that moving from a cold place to a warm place is more effective than antidepressants? Now imagine how this could help employers, governments, pharmaceutical companies and everyone else.

Imagine an unsupervised AI system that could remember all of your health and environmental data forever and link it to the data for your parents, your siblings and your offspring. It could continuously provide feedback based on new data and new studies while reducing the need for doctors to ask patients about their family history. Their family history would already be analyzed as part of a personalized health algorithm – not just for one generation but for many generations. Think of it as like a family tree on steroids.

With all of the technology available and everything around us being connected to the internet, real world data relating to patients' behavior and environment can be incorporated into disease prevention and patient care. Imagine the day when a patient shows up at the doctor's office and instead of the doctor asking "how do you feel?" and "have you been active?", they could already have this information available and displayed on a screen to educate the patient and to plan treatment around it. The patient could be involved and encouraged to track improvements in care over time using real world data. After all, real healthcare takes place outside of the hospital and doctors' office – and not when the patient shows up with a complaint.

MEET TITHONUS

There's a fantastic section in Michio Kaku's *Physics of the*

Future[176] where he tells the story of Tithonus, who appears in both Greek and Roman mythology.

When Eos, the goddess of the dawn, fell in love with handsome mortal Tithonus, she knew that he would eventually age and fade away. Eos was immortal and Tithonus was not, so Eos begged Zeus, the father of the gods, to give the gift of eternal life to Tithonus so that they could spend the rest of eternity together. Zeus took pity on them and granted Eos her wish.

Unfortunately, Eos forgot to ask for eternal youth. That meant that while Tithonus was able to live forever, his body continued to age, causing him to live for eternity in pain and suffering.

This cautionary tale is still relevant today. One of the big drivers of healthcare costs is that people are living longer due to advancements in medicine while living with chronic illnesses. These illnesses have to be managed – at great cost to the system.

Luckily, with new technologies, data, patient education, healthier living and a focus on prevention, intervention and cure, we can stop the legend of Tithonus from becoming a reality for us all.

TELEHEALTH

According to a 2017 Accenture report, "While clinical labor is in the early stages of moving to a virtual model, clinical interactions may transition more quickly. Certain clinical interactions will become virtual interactions that eliminate the need for an in-person visit. Care will no longer be based solely on physical location. Specialists will be called upon to treat a condition, regardless of geography."[177]

This will be made possible by telehealth, a rising trend in the healthcare industry in which patients are able to access expertise

---

[176] See: http://bit.ly/futurephysics

[177] See: http://bit.ly/newwaysofworking2017

through their smartphones. Imagine if instead of booking an appointment and having to take time out of your day to attend it, you could simply Skype a physician and get a virtual appointment. Because geography would no longer be a limiting factor, minor appointments could be dealt with by physicians from anywhere in the country if they're experiencing a lull at their actual facility, helping to spread the load.

The Accenture report notes that doctors could "clock-in at any time, making themselves available to patients. They [could] increase demand for their services to grow their practice and even access resources, such as telehealth training, clinical guidelines, peer support and billing."

Some providers are already starting to see success from telehealth programs. Emergency room wait times at New York's Weill Cornell Medical Center have been cut by more than half because of its Telehealth Express Care service.

It essentially works by streamlining the workflow, allowing patients to be removed from the main emergency room so that staff can focus on the patients who need the most help. The service is provided by the same doctors who work at the hospital and the processes are exactly the same, which means that there's no loss of quality in the service. Meanwhile, wait times for Telehealth Express Care patients have dropped from two-and-a-half hours to just 35 minutes.[178]

Another great example of telehealth in action comes to us via LiveHealth Online, an app which is able to connect moms with lactation consultants from the comfort of their own home. Lauren Majors, who works as part of a team of international board-certified lactation consultants, said, "I think it's incredibly convenient, particularly when you're immediately postpartum. There are so many doctor appointments to go to and you're incredibly fatigued. The baby is crying and needs feeding. It's great for a mother to get

---

[178] See: http://bit.ly/telehealthexpress

accessible care from her home."[179]

According to a study from research firm IHS Markit, American GPs will conduct 5.4 million video consultations per year by 2020. Meanwhile, in the United Kingdom, a paid service called Push Doctor has been launched which offers an online appointment "almost immediately" for £20 ($24). By reducing the time that they spend on administrative duties, the firm's physicians spend 93% of their time with patients compared to only 61% in Britain's public sector.[180] 61% might sound low, but it's huge compared to the 27% of their time that US physicians spend with their patients.[181]

Telehealth also has plenty of potential when it comes to the field of dentistry. According to Miles O'Brien, Science Correspondent for PBS NewsHour, "There are nearly one million Americans who visit the emergency each year because of dental pain at a cost that runs into the hundreds of millions."[182] Telemedicine could help to reduce this cost because the patient could simply use the camera on their device for a quick consultation and the dentist could prescribe medication. This would offer huge cost savings for hospitals, huge convenience for the patient and act as a shining example of preventative care by tackling the issue before it becomes too much of a problem.

THE BENEFITS OF TELEHEALTH

Telehealth has numerous benefits for every stakeholder in the healthcare industry, and it will only become more powerful as we have access to more and more patient-generated data from wearables, sensors and other new technologies. That means that

---

179 See: http://bit.ly/telehealthmoms

180 See: http://bit.ly/digitalhealthrevolution

181 See: http://bit.ly/doctorspatienttime

182 See: http://bit.ly/telehealthstat

providers will be able to access key information such as your vital signs remotely, enriching the video chats by providing key data.

A 2014 study by the Richard L. Roudebush VA Medical Center in Indianapolis found a 96% overall satisfaction rate among veterans who used the center's telemedicine platform. Veterans travelled almost 800,000 fewer miles in the two-year span and it saved the hospital over $330,000.[183]

Another big benefit of telehealth is its potential to bring better quality healthcare to rural or isolated communities. If there's no doctor available, or if the local physician doesn't have the expertise that they need, then a specialist can be consulted through a telehealth system and a better standard of healthcare can be offered to people who might not otherwise have had access to it.

You can see this in action already through innovations like the Cardio Pad®, the brainchild of 26-year-old Cameroon-based engineer Arthur Zang. Zang started work on the device in 2009 while studying engineering at the Polytechnic High School of the Yaoundé I University, Cameroon. The device allows healthcare workers in rural areas to send the results of cardiovascular tests to specialists using a mobile phone, profoundly increasing the availability of cardiac healthcare services for people living in remote, rural areas of developing nations. This can make a huge difference to patient outcomes, saving lives in countries with inadequate health infrastructures and understaffed and underfunded health systems.

Zang's application was tested and compared to others using the MIT MIB database with an impressive 98% success rate when it comes to digital cardiac signal processing. A prototype of the tool has been available since 2010, with efforts since then focusing on refining and marketing the tool. Zang has been recognized with the Rolex Award for Enterprise and the Africa Prize for Engineering Innovation for his invention, which is said to be Africa's first medical computer tablet.

But while Zang and his team have focused on increasing the

---

[183] See: http://bit.ly/telemedicinestudy

amount of face time that patients spend with specialists, new technologies can also increase the amount of data that these specialists have access to. Let's say you spend an average of ten minutes a month in a visit with your doctor. That leaves 43,790 minutes a month that you're elsewhere. Those occasional visits to the doctor provide an incomplete picture of your health. Luckily, the advent of new technology and new data sets will combat the problem, resulting in more data than ever before for providers to base decisions on.

Social factors are a component in one out of three deaths. Social isolation can increase the risk of heart disease by 29% and the risk of stroke by 32%. And between 75-90% of primary care visits are due to the effects of stress.[184] These statistics alone should be enough to show how doctors are currently unable to track each of the different factors that they need to know about their patients. A person's neighborhood could have more of an impact on their long-term prospects than any medication could, and a value-based healthcare system will put this information front and center. Physicians can't make informed decisions unless they have access to data.

This data belongs to us as the patient, rather than to the doctors. But despite this, it can also help medical care as a whole by providing both specific data about you as an individual and anonymous data that can be aggregated to develop new treatments and expand our medical knowledge.

But this creates another problem. Howard Green MD captured the quandary in an article in which he wrote, "Why people admire the value of a body left to science to be examined retrospectively, yet withhold interoperable and integrated medical chart data from analysis in real time, remains the great mystery currently preventing the advancement of clinical science and the betterment of our health."[185]

The health insurance, pharmaceutical and EHR industries have

---

[184] See: http://bit.ly/socialdeterminantshealth

[185] See: http://bit.ly/greenmdarticle

all recognized and acted upon the value of our personal medical data. These industries are already using our data to stretch their profits, and yet people are refusing to share anonymized and deidentified medical data which could be analyzed by researchers, doctors and artificial intelligence systems.

If patients continue to refuse to share their data, it severely limits the ability for the healthcare industry to learn from them and to improve outcomes for all. It's my hope that this book will help to convince people of the importance of sharing their data. We all have a part to play in the future of healthcare.

ZNA IS THE NEW DNA

In January 2015, then-president Barack Obama launched the Precision Medicine Initiative (PMI), a revolutionary plan that aimed to rethink our approach to health and medical treatment by taking advantage of advances in genomics, emerging methods for managing and storing data and the boom in health information technology. When she announced the 2016 budget, Sylvia M. Burwell, the Secretary of the US Department of Health and Human Services (DHHS), announced that $215 million would be allocated to the PMI, with $200 million of it set aside to launch the All of Us program, "a national cohort of a million or more Americans who volunteer to share genetic, clinical and other data to improve research."[186]

But it's naïve to think that precision medicine only goes gene-deep. The truth is that we're affected by a huge number of factors, from where we live to the air that we breathe, the water we drink and the environment around us. This is underscored by National Institutes of Health (NIH) director Dr. Francis S. Collins, who tweeted, "If DNA is our biological blueprint, ZNA (ZIP code at

---

[186] See: http://bit.ly/allofusprogram

birth) is the blueprint for our behavioral and psychosocial makeup."[187]

The truth is that our health is determined by all sorts of social determinants, including factors like socioeconomic status, education level, the physical environment, employment status and social support networks, as well as the amount of access that's available to healthcare services. The social environment encompasses factors like the crime rate and the affluence of the neighborhood, while when I talk about the physical environment, I'm talking about whether your neighborhood is designed to be walkable, whether it's home to heavy industries and whether it has access to healthy foods or whether it's littered with fast food restaurants and liquor stores.

Drawing a parallel between these social determinants and overall health is hardly revolutionary. In fact, studies have shown that social factors can have a more pronounced effect on health than the distinct biological differences between people. Income and educational achievement are as strongly associated with hypoglycemia risk in diabetes patients as known clinical risk factors.[188] Meanwhile, people who live in neighborhoods with limited healthy eating and physical activity resources have a higher risk of being diagnosed with type 2 diabetes mellitus.[189]

In a recent keynote address about the links between segregation and poor health at an event at the Dana-Farber Cancer Institute, Dr. Melody Goodman echoed the words of NIH director Francis Collins by explaining, "Your zip code is a better predictor of your health than your genetic code."

This is no overstatement. In fact, if you open your eyes to it, you can see it in action. Take Delmar Boulevard, a major east/west street in St. Louis, Missouri, which marks a sharp dividing line between the poor, predominantly African American neighborhood to the north and the more affluent, predominantly white neighborhood to

---

187 See: http://bit.ly/franciscollinsquote

188 See: http://bit.ly/riskfactorsdiabetes

189 See: http://bit.ly/type2diabetesfactors

the south. Education, health, health literacy and access to resources also follow the "Delmar Divide", with residents to the north less likely to have a college degree and more likely to have cancer or cardiovascular diseases.[190]

There are dramatic differences in life expectancy across the United States depending upon where you're born. The lowest longevity places tend to be in the south, while the highest are in the northeast and the west. The same applies around the world, with dramatically different life expectancies between developed and developing countries. Therefore, while genetics and healthcare are critical, ZIP code, race, education and class trump genetics as predictors of health – which means that understanding how our environment and conditions shape our health could lead to improved treatments and better preventative care.

In an article for The Conversation, physician Brian Schwartz and chronic disease epidemiologist Annemarie Hirsch argued, "Technological advances in geographic information systems (GIS), including digital software, maps and longitudinal data sets on socioeconomic and environmental factors, can be utilized to describe various aspects of an individual's community and environment. Just like we can sequence a person's DNA, we could use this data to sequence our 'ZNA' from birth to death."[191]

What does this mean for us? Well, in the not-so-distant future, as electronic health records (EHRs) become more common, it'll be possible – and should be compulsory – to link new data about local environments with patients' medical records. I believe that this link is vital for both clinical care and research, and perhaps most importantly, it will also allow an increased focus on prevention. This in turn will translate to decreased costs for the healthcare system – and that's good news for everyone.

A value-based healthcare system would put this information front and center. Physicians can't make informed decisions unless

---

[190] See: http://bit.ly/delmardivide

[191] See: http://bit.ly/znathenewdna

they have access to data, as well as the knowledge they need to make sense of it. At the moment, this socioeconomic and environmental data is siloed away from the patient and the clinician at the point of care, despite the fact that it could help medical care by providing both specific data about you as an individual and anonymous data about the population as a whole. This data could enable us to develop new treatments and to expand our medical knowledge.

THE SMARTPHONE AS THE NEW PHYSICIAN

The increasing availability of smartphones and tablet computers – as well as their ever-increasing processing power and functionality – is ushering in an era in which we carry everything we need to live a healthy lifestyle right there in our pockets.

We can see this happening at a basic level already. One company in Scotland has launched a free smartphone app that's designed to increase users' understanding of prescription medication.[192] It's estimated that up to 50% of patients prescribed medication by their doctor fail to take them at the right time and at the right dosages, costing the NHS up to £300 million per year. The app works by reminding patients to take their medication and explaining how it actually works. All you need to do is point your phone's camera at the medication's barcode and the application does the rest. Meanwhile, pharmaceutical companies can access information on the effects that their products have in the real world, and this all happens without any personally identifying information being either logged or shared.

Of course, the potential for smartphones to be used as healthcare devices is largely limited by manufacturer uptake. Nevertheless, major companies are already taking steps in the right direction, and Apple has even filed a patent that describes how the iPhone could be

---

[192] See: http://bit.ly/medicationapp

used as a health sensor to detect "blood pressure index, blood hydration, body fat content, oxygen saturation, pulse rate, perfusion index, electrocardiogram, photoplethysmogram, and/or any other such health data."[193]

This isn't a science fiction novel and so I'm not going to go out on a limb and make predictions. Still, while patents aren't always a reliable source of rumors and information, it's hardly the first time that a major device manufacturer has expressed an interest in healthcare. It'll be interesting to see what the future brings.

PARTICIPATORY MEDICINE

Dr. Danny Sands, the co-founder and chairman of the Society for Participatory Medicine, says, "For too long, both patients and healthcare professionals have thought of healthcare as a car wash, with the patient passively moving through the healthcare system car wash, getting health sprinkled on them and coming out healthy. This lack of engagement results in dissatisfaction, high costs and poor quality care. We need to reimagine healthcare as an active collaboration between the patient and the healthcare professional."[194]

As you can see, the main idea behind participatory medicine is patient involvement and active collaboration. In other words, transforming the culture of healthcare from transactional to value-based. They do this through everything from research and education to advocacy and community groups, empowering patients to take control over their own bodies.

Every initiative for positive change in healthcare has to achieve three main goals: improved health outcomes, greater satisfaction and lower costs. Participatory medicine ticks all three of those boxes, and the devices and technologies that we've talked about will help to

---

[193] See: http://bit.ly/applehealthpatent

[194] See: http://bit.ly/participitatory

make it a reality.

We no longer live in a world where the doctor makes the sole decision. Patients and caregivers want to be educated and empowered so that they're able to take part in the decision-making process. They'll no longer tell the doctor to just "do what you think is best" because they're out of their depth and they see them as some magician who can give them a pill and fix them. They'll take ownership of their own bodies, understand the options that are available to them and make an informed choice about the best route to take towards the future.

OUR HEALTH IN OUR HANDS

This is the title of a May 2014 Wired article in which Daniel Kraft, faculty chair for medicine and neuroscience at Singularity University, argued, "The healthcare world that most of us experience – and the one that clinicians are traditionally incentivized to operate in – has been one of 'sick care', in which we focus our time and energies on treating diseases once they have appeared, or reached a point where they can no longer be ignored."[195]

But as Kraft points out in his article, that paradigm is starting to change. New healthcare challenges like obesity and the ageing population are compounded by a lack of primary care physicians. Fortunately, new, interconnected technologies and devices are set to turn the internet of things into the internet of the body.

How will it do this? Well, Kraft suggests that the data from these devices could be shared through APIs to connect data to the cloud. This data could then be analyzed and changes could be "prescribed" for "improving wellness, diagnosis and therapy". Kraft says, "Imagine a GPS system for your health. It knows your habits, your genomics and your goals and can help you reach a target, whether

[195] See: http://bit.ly/healthinhands

that be to run a marathon, lose weight, manage hypertension or lower your risks for cancer."

The result is a society in which our health really is in our hands. This prediction was backed by a 2014 PharmaVOICE article that explains, "Patients have the power of digital technology to fuel their quest for better health, which is forcing the industry to keep pace with their digital needs."[196]

The bottom line is this: while pharmaceutical companies struggle with the adoption of technology due to restrictive regulatory processes, they really have no choice. With patients increasingly demanding more control over their health – and with device manufacturers more than happy to cater to that demand – these devices are coming whether we like it or not.

As Richard Nordstrom of Liberate Health says, "Pharma has a responsibility to place digital tools and content into the hands of clinicians, which will help improve patient health outcomes." Outcome is the name of the game. If patients can take their health into their own hands to improve those outcomes then so much the better.

THE PATIENT'S VOICE IS MISSING

While I was researching this chapter, I spotted a fantastic piece by Dr. Aniruddha Malpani about how the patient's voice is missing from EMRs. Malpani explains, "What typically happens during a medical consultation is that the doctor asks the patient a series of structured questions and then uses his medical expertise to convert the patient's story into a format which doctors have agreed is the best way to document the patient's medical journey. Now, all this is well and good when you have an intelligent doctor, but often the patient's story [gets] distorted when it is seen through the doctor's

---

[196] See: http://bit.ly/pharmavoicetech

prism. He will often try to fit square pegs into round holes and selectively record what he feels is medically relevant. This often means that the medical record is incomplete and inaccurate and can end up doing more harm than good."[197]

What's interesting about this is that Malpani agrees with Joe Biden, who we talked about back in chapter three. You'll recall that when Biden talked about patients having access to their medical records, the CEO of Epic asked him why he'd need access to the data if he couldn't understand it. It seems to me that if medical records are too difficult for patients to understand, that's a problem with the medical records and not with the patients.

Malpani points out that "once an incorrect fact or interpretation is entered into the EMR, it becomes gospel truth and it's very hard to correct it later on." But patients know themselves better than any doctor, so why not give them the opportunity to correct the mistakes they see? After all, when we start to aggregate this data to monitor overall population health, inaccurate EMRs on an individual level could lead to us drawing false conclusions about the population as a whole. If doctors are subconsciously over-diagnosing Lyme disease because it's been in the news, that could make us think that Lyme disease is on the rise when no such phenomenon is taking place. Even in the future of healthcare there'll be room for human error.

"It's often because the patient's story gets distorted when doctors retell it that so many medical errors occur," Malpani explains. "This is why senior doctors will first talk to the patient and form an independent assessment before reading the medical notes, because they know that they can distort their perception. Why do we continue to infantilize patients and treat them as ignorant children, from whom their medical notes should be hidden? It's ironic that today if the patient wants access to his medical records, he has to get hold of a lawyer to bully the hospital [into giving] him a copy."

Ultimately, the best solution is likely to be one in which the patient's voice is represented by a set of notes on their file. I'm not

---

[197] See: http://bit.ly/patientvoicemissing

saying patients should be able to go in and change the information without the oversight of a qualified physician, but I am saying that allowances should be made to incorporate the patient's voice into their health record. Otherwise patients will become alienated, and they won't be able to play a role in their own treatment through participatory medicine unless they understand what they're doing and why they're doing it.

"The patient gets frustrated trying to give the new doctors he sees the correct version of events," Malpani explains, "but sadly no one listens to him because when the medical record differs from what the patient says, the medical staff trusts the medical record, rather than the poor patient. This is ironic because often [the] patient's version is much more accurate – after all, he knows what's happening to him! This would also be a great way of checking the quality of the medical record and checking that the information it contains is accurate. Because the patient is so deeply invested in his own medical care, he will make sure that the record is updated, accurate, and that there are no inconsistencies. This openness and transparency will increase the confidence and trust patients have in the healthcare system. Patients are the largest untapped healthcare resource, and we need to give them the right to add their own voice to their own records."

Remember, we're all stakeholders in the future of healthcare, and one day we'll all inevitably be patients. It's time for us to make a stand and to say "this can't go on." The patient's voice is more important than ever.

THE IMPORTANCE OF SENSORS

I agree with The Human API and their claim that "our cars are better managers of their health than we humans are."[198] If we're not

---

[198] See: http://bit.ly/humanapi

careful then this disparity will only get worse over time as self-driving cars are developed that use thousands of 'listening' and 'seeing' devices to know the status of everything from the traffic along the route to the temperature inside the car and the amount of pressure inside the tires.

These vehicles are even being programmed in such a manner that they know when they need a checkup and automatically take themselves to service centers. By doing this, they can stop a catastrophic fault from taking place and prolong the life of their components. If this sounds familiar it's because it's exactly the same principle as we see in the future of healthcare when we talk about stopping illnesses and diseases in their tracks instead of waiting for them to manifest themselves.

The Human API is a project with the noble goal of providing "lab-in-a-box capabilities to be able to test at home and not rely on doctors to run tests." The idea is that we need to take care of our own health in the same way that autonomous cars look after their engines. To do this, we need sensors that are able to monitor us around the clock, from wearable devices like the Fitbit to smart home devices which can help you to live more healthily by monitoring how often you raid the fridge or go outside for a cigarette.

There'll be skeptics along the way of course, but there always is. It's like Arthur Schopenhauer said: "All truth passes through three stages. First, it is ridiculed. Second, it is violently opposed. Third, it is accepted as being self-evident."

The future of healthcare is on its way. And the good news is that we're already somewhere between steps two and three.

THE IMPACT OF TECHNOLOGY

Ultimately, technology is set to have such a huge impact on our lives that it's impossible to summarize it in a single book. Still, I wanted to round this chapter off by sharing a few examples of the revolutionary tech that's set to be ushered in with the future of healthcare.

One of the big growth areas is likely to be that of 3D printed drugs. The FDA has already approved the first of them, an epilepsy drug called Spritam that's printed out layer by layer to make it dissolve faster than average pills. Meanwhile, scientists from the University College of London have tested 3D printing drugs in shapes like dinosaurs and octopuses to encourage medication adherence amongst children.[199]

The fact that Amazon is getting into the pharmaceutical game is a good indicator of the direction things are likely to move in. Eventually, they'll be able to bypass hospitals and insurers and go directly to the consumer, 3D printing medication on demand and delivering it to your door. Better still, they can base those drugs on your own unique gene expression and medical history through the use of personalized medicine. Big data makes all of that possible.

Meanwhile, new innovations like artificial pancreases are being brought to market for use amongst diabetes patients. They're designed to constantly measure blood glucose levels and to administer insulin and glucagon automatically, dramatically cutting down on incidents where diabetes patients suffer an attack during the night and can't be woken up by the alarms on traditional monitors.[200]

Other intriguing uses of new technology include a project to use Alexa in clinical trials. The idea is to get it to talk to patients to remind them to take their medication and then to measure the impact that it has. Orbita, a company which develops voice-first software for connected home healthcare devices, recently teamed up with clinical trial data firm ERT to "create interactive voice surveys and collect voice responses" and to provide "built-in analytics to track user engagement and respond to user input." Harry Wang, senior director of research at Parks Associates, an internet of things

---

[199] See: http://bit.ly/printedpills

[200] See: http://bit.ly/artificialpancreasmesko

market research consultancy, said, "Voice-based UI innovations have huge implications for the healthcare industry, particularly in areas where patient participation, interaction and engagement is critical to the market success of digital health solutions and services. This collaboration between Orbita and ERT is a clear example of voice recognition technology's potential in healthcare and we expect many more use cases that incorporate voice UI to emerge."[201]

Even Nestlé is getting into the game. The food and drink company is perhaps best known for brands like Nescafé, Kit Kat, Smarties and Nesquik, but they're now billing themselves as "the world's leading nutrition, health and wellness company." They've been trying to develop medical foods for patients with various diseases, as well as foods that are designed to help people heal more quickly after surgery.

This is the cutting edge of targeted healthcare and has been referred to by many as "the wild west of medicine". A meal that nourishes the body but that's also specifically designed to treat Alzheimer's would be a huge step forward. Medical foods don't have to meet the same requirements as prescription drugs, but the research costs alone will be enough to slow down development. Luckily, food manufacturers can work with AI to identify patient needs and to speed up the process.

Before we move on to take a look at disruption for the pharmaceutical industry, I'd like to highlight one more company that's helping to usher in the future of healthcare. That company is called 23andme, and they're huge innovators in the targeted medicine industry because they distribute home genetic testing kits.[202] For just $200, you can get a report on every chromosome in your genetic makeup and specific recommendations that are based upon your unique genetic structure. Like opening up a book, you can read your DNA to arrive at insights about your health, your personal traits and even your ancestry. This is beneficial for you as a

---

[201] See: http://bit.ly/orbitaert

[202] See: http://bit.ly/23andmesite

patient, but it's also a boon to the companies which will use that data to better understand diseases.

Eventually, pharmaceutical companies will be able to use this data to create personalized drugs for patients. That can only be good news for all of us.

CHAPTER NINE: THE FUTURE FOR BIG PHARMA

"MEDICINE IS NOT ONLY A SCIENCE; IT IS ALSO AN ART. IT DOES NOT CONSIST OF COMPOUNDING PILLS AND PLASTERS; IT DEALS WITH THE VERY PROCESSES OF LIFE, WHICH MUST BE UNDERSTOOD BEFORE THEY MAY BE GUIDED."

- PARACELSUS, PIONEERING SWISS PHYSICIAN

WATCH OUT PHARMACIES AND BIG PHARMA: disruption is coming.

Thanks to the rise of new technologies, it's no longer pharmaceutical companies versus other pharmaceutical companies. Instead, they're facing increased competition from platforms like 23andme and patients like you and I. For pharmaceutical companies to survive, they'll need to adopt a new business model which incorporates the use of technology in order to adapt to the future healthcare market. This means that the pharmaceutical companies of the future may more closely resemble today's tech giants than their contemporary equivalents.

There's a great piece by Niven Narain in Drug Discovery & Development magazine in which he talks about just how overdue the pharmaceutical industry is when it comes to disruption.[203]

---

[203] See: http://bit.ly/overduedisruption

Narain explains, "Pharma has lagged behind when it comes to adopting and implementing new technological advancements such as artificial intelligence or machine learning. This has provided an opening in the pharma space that tech companies are looking to fill. Apple, Intel and Google have all recently made large investments buying up AI startups. And while none of them have experience in drug development, they're all aggressively positioning themselves to enter the healthcare marketspace. Silicon Valley is accustomed to rapidly innovating and embracing new technologies and could very well surpass traditional pharma to disrupt the whole industry."

Narain seems to agree with a concept that I've already discussed elsewhere. Pharmaceutical companies of the future can adopt a new business model in which they charge a premium for precision medicine instead of adopting a one-size-fits-all approach.

This is a concept that some innovative big pharma companies have embraced. However, for the late adopters, the future is still coming sooner than later. As Narain explains, "An approved drug may only benefit 10% of patients, yet it's prescribed to the entire patient population and is reimbursed by health plans irrespective of its actual clinical benefit. The world of big business, which is driven by the need for profits, can lose sight of the goal of helping to improve the lives of patients in need. But by disrupting the industry through these emerging technologies, we can deliver treatments to patients in a precision way, improving outcomes while also reducing overall healthcare costs."

PERSONALIZED MEDICINE

One of the big trends for pharmaceutical companies will be the shift towards personalized medicine. We'll no longer work with one-size-fits-all drug treatments. Instead, pharmaceutical companies will learn to tailor their medicines based on different use cases.

We're already starting to see this happen. The FDA recently approved Keytruda, an innovative new immunotherapy treatment which is deployed based on the genetic characteristics of tumors instead of their location on the body.[204] This big breakthrough was heralded as a landmark for personalized medicine, and it's believed that 4% of all advanced cancers harbor the genetic abnormalities that Keytruda is designed to treat.

We've talked elsewhere about the inherent advantages of personalized medicine, and I'm not going to go back over old ground by preaching to the converted. By now, you'll have seen how personalized medicine can help to intercept illnesses before they become a problem and lead to dramatically improved outcomes for patients.

One thing that we haven't considered, though, is the impact that this will have on pharmaceutical companies. The shift towards personalized healthcare will require the industry to rethink its approach to doing business and to adopt a new business model that puts personalized medicine at the heart of it. If this means making less medication or switching their focus to developing technological solutions, so be it.

HOW TECHNOLOGY WILL CHANGE DRUG DISCOVERY

Benjamin Carrington, Senior Program Director of Biomarkers, Diagnostics and Artificial Intelligence at Hanson Wade, put it best in an article he shared on LinkedIn.[205] Carrington said, "There's such a breadth in the potential verticals within a pharmaceutical company where AI can provide benefit, from early molecular drug design, scientific research mining and drug repurposing to clinical trial enhancement, patient recruitment and selection, sales and marketing and business intelligence."

Carrington is backed up by Boehringer Ingelheim and their chief

---

[204] See: http://bit.ly/cancergeneticsdrug

[205] See: http://bit.ly/carringtonarticle

data scientist, Philipp Diesinger. With over 130 years of history and 45,000 employees, Boehringer is well established as one of the world's top twenty pharmaceutical companies, but that doesn't mean that they're reluctant to adopt new technologies. In fact, Diesinger says, "The entire industry is looking at data science and AI."

For Boehringer, "the combination of AI, big data and new perceptions of these deep analytical methods" is able to reduce the cost of drug development and enable earlier decisions on potential pipeline candidates. Diesinger says the company wants to evolve from a pharmaceutical company to a holistic healthcare company with the help of AI, noting that AI has already transformed the financial industry "using theoretical physicists and mathematicians to optimize training."[206]

When it comes to AI and machine learning, the greatest potential for the pharmaceutical industry is the way in which it can redefine clinical trials. AI can run simulations to identify different areas of potential research, it can crunch the numbers to identify the patients you ought to work with and it can even simulate different drug combinations to help you to identify potential clashes or side effects.

It's able to do this using 'population health' data, which aggregates information about a community as a whole to draw conclusions. As Howard Green MD explains, "We have the ability in real time to compress, tabulate and reveal 200 years' worth of epidemiological and charted patient health information in seconds."[207]

AI can't do all of the work for you, but it can speed up the process of drug development and make it more affordable for providers to carry out clinical trials and research new treatments. It's about time, too. After all, it costs over $2.5 billion to develop a new drug that gains market approval. If AI can cut that cost, there'll be no choice in the matter.

---

[206] See: http://bit.ly/aitransformation

[207] See: http://bit.ly/populationhealthdata

UNDERSTANDING HOW PATIENTS MAKE DECISIONS

Patients are increasingly taking their health into their own hands. If they start to feel sick, they Google the symptoms or call a helpline to ask for advice. They like to carry out a little research before they visit a doctor, and this is especially true for relatively minor illnesses and injuries. After all, they can often find the information right there and then instead of having to wait for an appointment and heading across town to meet their doctor.

In fact, one in every twenty Google searches is now for health-related information.[208] The average person makes around 540 searches per year[209], which means we search for health-related information every two weeks on average – which is much more often than we see our physicians.

On top of that, people are increasingly paying attention to online reviews and information from their peers. 74% of consumers identify word-of-mouth as a key influencer in their purchasing decisions[210] – and if you think that their healthcare decisions don't count then think again.

These days, consumers have the power. Pharmaceutical companies are starting to realize this and to see that their business model – and the industry as a whole – is changing. It's an exciting time to be alive, and the pharma companies that survive will be those who see the shift as an opportunity and not a threat.

USING GENETICS

Nobody likes wastage, so why should pharmaceutical

---

[208] See: http://bit.ly/googlestatshealth
[209] See: http://bit.ly/searchesperyear
[210] See: http://bit.ly/wordofmouthstats

companies be any different? Genetic profiling and personalized medicine could allow us to prescribe drugs more effectively, avoiding the wastage (in terms of time, money and resources) that occurs when patients are mis-medicated.

Genetic testing isn't difficult (and it's not even prohibitively expensive), and there a number of drugs on the market that already have an FDA label for DNA data. But the infrastructure which could make genetic testing as simple as swabbing your mouth at the doctor's office is yet to see widespread adoption. Mixing genetic data with other data could improve the efficacy of medication and lead to a more personalized approach to healthcare. More importantly, it could highlight potential adverse effects before they become a problem.

Kaiser Permanente is just one of many companies that are disrupting the pharmaceutical industry by better studying drugs and their impact. They've successfully used big data to study the incidence of blood clots within a group of women taking oral contraceptives, discovering that one drug increased the threat of blood clots by 77%.

With patients consuming so much medication in a single lifetime, it's almost impossible to predict the subtle interplays of the different drugs they've taken. This becomes infinitely more difficult when you start to consider lifestyle factors and genetic predisposition, which is why big data and machine learning is needed to make sense of it all.

OUTCOME-BASED PRICING

Novartis AG has already showed the potential for big pharma to revolutionize the industry with its CAR-T cell therapy, which was recently approved by the Food and Drug Administration. It's the first gene therapy to be available in the US and was approved for young people up to the age of 25 with a form of acute lymphoblastic leukemia (ALL).[211]

But what's interesting about this treatment isn't the fact that it's the first gene therapy, and nor is it the fact that it's listed at $475,000 for one-time treatment. It's the fact that the cost of the therapy will be "based on the clinical outcomes achieved" and will specifically only allow for payment if all patients "respond to [the drug] by the end of the first month."

This is good news for patients and good news for the industry, too. For pharmaceutical companies to charge a premium under the new model of healthcare, they'll need to supply the right medication to the right patient at the right time. Adopting a system in which they're only paid for successful outcomes can only be a step in the right direction. As I mentioned previously, the name of the game is outcomes.

THE FUTURE OF PHARMA

London's Royal Pharmaceutical Society recently published a report on future models of care[212] which delivered a number of recommendations that are worth some attention from pharmaceutical companies. Recommendations for pharmacists included:

Focusing less on the distribution of medication and more on providing a range of services.

Helping people to get the most from their medication and focusing on keeping them healthy instead of

---

[211] See: http://bit.ly/outcomebasedpricing

[212] See: http://bit.ly/rpsreportpharma

treating symptoms.

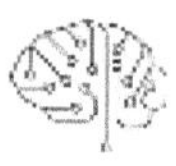
Taking the initiative and driving change and innovation at a smaller scale instead of waiting for sweeping national reform.

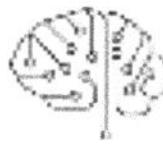
Collaborating with each other across community, social, secondary and tertiary care – as well as with other healthcare professionals.

It's clear, then, that the very business model for massive pharmaceutical companies is under threat, but the opportunities are huge for the companies that are able to adapt to it. It's a case of adapt or die, because if existing pharmaceutical companies move too slowly then they'll quickly be overtaken by their more agile upstart competitors.

In fact, pharmaceutical companies could find themselves under threat from the patients themselves. Companies like Kaggle have come along and changed the way that we look at big data analytics. Instead of using some proprietary algorithm to make sense of the data, they use crowdsourcing and gamification to bring "citizen scientists" together to solve problems. The company has even figured out how to make a profit while simultaneously giving cash prizes out for people who successfully make sense of big data, and they apply this approach to healthcare and scientific research, as well as to big business. For example, they've helped in the search for dark matter and competed to predict HIV progression with a limited dataset[213] – all from their bedrooms as a hobby.

Pharmaceutical companies could find themselves in trouble from citizen scientists and entrepreneurs who are able to disrupt the healthcare industry at the rapid rate that new technologies require. One industry commenter went so far as to say: "I believe that drug producers should have an advisory board including patients who

---

[213] See: http://bit.ly/hivprogression

have experience with the given company's products. It would be easier to develop new products if the exact needs of the customers are well-known. Only with their help would it become possible to create a healthcare system that is futuristic even decades after the first plans were drawn."[214]

---

[214] See: http://bit.ly/toptentrends

CHAPTER TEN: THE FUTURE FOR PHYSICIANS

"I BELIEVE IN A FUTURE WHERE THE VALUE OF YOUR WORK IS NOT DETERMINED BY THE SIZE OF YOUR PAYCHECK, BUT BY THE AMOUNT OF HAPPINESS YOU SPREAD AND THE AMOUNT OF MEANING YOU GIVE. I BELIEVE IN A FUTURE WHERE THE POINT OF EDUCATION IS NOT TO PREPARE YOU FOR ANOTHER USELESS JOB, BUT FOR A LIFE WELL LIVED. I BELIEVE IN A FUTURE WHERE JOBS ARE FOR ROBOTS AND LIFE IS FOR PEOPLE."

– RUTGER BREGMAN, AUTHOR OF UTOPIA FOR REALISTS

WHEN I TALK ABOUT the future of healthcare, many of my colleagues in the medical field get anxious. They worry that they're not going to understand the technology, or even that robots and computer programs could replace them altogether.

It's certainly true that many jobs are under threat from automation. One study from Oxford University and Deloitte found that 35% of British jobs were at high risk of computerization in the next two decades – but that the country's 232,000 medical practitioners had one of the lowest likelihoods of automation.[215]

Far from replacing humans, intelligent new technologies will

---

[215] See: http://bit.ly/automationcalculator

work with them to provide a better prognosis for their patients. Virtual assistants, internet-connected technology and surgical robots will have a presence in the hospital, but they'll be viewed in the same way that we look at X-Rays and MRI scanners. They're tools, not a threat to your job and your livelihood.

THE SCIENCE OF CHANGE

It's only natural for us to be afraid of new technology. After all, change can be terrifying and there's nothing worse than uncertainty. Psychologist Harriet Lerner, the author of The Dance of Anger, says, "We don't resist change because we're neurotic or cowards. The will to change and the desire to maintain sameness coexist for good reasons, and they're both essential to our emotional health and to the continuity of our identity."[216]

Put simply, medical practitioners *want* the future of healthcare, but they don't want to take responsibility for making it happen. There's an internal conflict between the drive to provide better healthcare and the struggle to simply stay on top of their existing workload. But change happens inevitably over time, and many forward-thinking doctors are not just embracing technology – they're advocating for it on their patients' behalf.

The physicians of the future will use technology as a matter of course. Better still, they won't just use it to help their patients – they'll also use it to help themselves. Right now, though, there are two different types of practitioner – there are those who are already embracing the future, and there are those who are trying to fight it and who will ultimately be overwhelmed. The future is coming whether we like it or not, and the disruption in the healthcare industry will change the way that physicians work for the better.

As Dr. Warner Slack said: "Any doctor who *could* be replaced by

---

[216] See: http://bit.ly/scienceofuncertainty

a computer *should* be."

FILLING IN THE GAPS

The average patient with five chronic conditions spends only 15 out of the 8,760 hours in a year in front of a doctor.[217] That leaves 8,745 hours unaccounted for.

In the era of value-based healthcare, physicians will look increasingly to internet-connected devices to analyze their patients between meetings. If the patient is wearing a smartwatch or using another device to monitor themselves around the clock, physicians will have access to all 8,760 hours of data and will be able to monitor conditions in real-time. The patient may only need to see them for 10 hours instead of 15, and yet they'll be given a much greater standard of healthcare which focusses more on prevention than on delivering a cure.

This technology already exists, at least to a certain extent. Baidu's VP & Chief Scientist Andrew Ng and his team at Stanford University have created a machine-learning model that can identify heart arrhythmias from an electrocardiogram better than an expert. And despite the kneejerk reaction from technophobes, Ng was positive about its reception, saying: "I've been encouraged by how quickly people are accepting the idea that deep learning can diagnose at an accuracy superior to doctors in select verticals."[218]

Ng's model took advantage of deep learning, drawing upon 30,000 30-second clips from patients with arrhythmia. These clips were provided by iRhythm, a company that makes portable ECG devices, which shows the potential for larger, private companies like Apple to get into the healthcare game. Deep learning involves using huge amounts of data and pumping it through a simulated neural

---

[217] See: http://bit.ly/connectedhealthnews

[218] See: http://bit.ly/machinesplaydoctor

network until it's able to accurately recognize problematic ECG scans. Machines are better than humans at this because they're equipped to spot complex patterns in images and audio at a high speed, which is why we already have image and voice recognition systems that are better than human beings.

Even if artificial intelligence had the same capabilities as a human when it came to spotting abnormalities, there are other factors to consider. The problem with hiring humans is that it's expensive and humans can't be there all the time. Technology can, and it can take care of the little things so that physicians are free to spend more time doing the jobs that artificial intelligence can't handle.

In many ways, automating tasks is the only way that healthcare can be sustainable, especially with the world's population set to reach 9.7 billion by 2050.[219] Patients will need to become their own doctors, at least to a certain extent. Wearable devices and new technologies will give people the power to play a role in their own healthcare, pushing towards a focus on prevention rather than the cure. With predictive analytics and all the data we'll have, diseases could be anticipated ahead of time and measures could be taken to address them before patients started showing symptoms. If we do this, we'll have a future where people live longer and healthier lives, and medication will be the last resort. It'll be taken later in life when we really need it.

THE FUTURE IS HERE

Some physicians are terrified by technology and unwilling to admit that the future of healthcare is inevitable. The truth is that humans and machines can work together for better outcomes – and that they're already doing so right now.

---

[219] See: http://bit.ly/worldpopulation2050

Healthcare is experiencing the dawn of the fourth industrial revolution and the possibilities of tech in healthcare are endless. Better still, there are already plenty of innovative companies on the market who are showing promise. For example, Seton Healthcare tried to reduce the number of readmissions for congestive heart failure through the smart use of technology. They applied predictive models to their analytics and then used the data to introduce early interventions for high-risk patients. This led to fewer visits to the hospital, a reduced mortality rate and improved care and quality of life for the patients.

Unfortunately, new technologies also come with their downsides. One report by Google's DeepMind Health, which is working with the NHS to apply AI and machine learning to their vast amounts of data, found that doctors were using Snapchat to send patient scans to each other.[220] This was described as "clearly an insecure, risky, and non-auditable way of operating" and echoes concerns about data privacy that are already holding the industry back from achieving its full potential. To their credit, the doctors aren't necessarily at fault – they saw the tools as allowing them to carry out their job more efficiently. But if the future of healthcare is to be ushered in, doctors must be trained in the correct use of new technologies and provided with the tools they need to use them. They shouldn't be using Snapchat. They should be given access to bespoke tools of their own.

MEASURING WHAT MATTERS

Let me introduce Paul Tang. Tang is a primary care physician, and our story begins when he visited his wife in the hospital after knee replacement surgery. Tang wanted to know when she'd be able to go about her day-to-day routines based on the surgeon's

---

[220] See: http://bit.ly/doctorsusingsnapchat

experience of previous patients, but the surgeon was unable to provide a definitive answer. That's when Tang realized that he didn't know – and that most physicians don't really know what happens to their patients once they leave the hospital.

But Tang, who is also the chief health transformation officer for IBM Watson's Health, didn't leave it at that. He realized that doctors were unable to measure what really mattered to their patients but that Watson might be able to help. In his wife's case, it could mine the data of similar patients to give an informed answer as to how long it would take her to recover.[221]

The sad fact is that many doctors have lost their way. Like the wannabe village vet who imagines a career tending kittens and puppies and ends up working for a factory farm, physicians often find that the reality of modern healthcare is at odds with their vision of helping patients to live longer, happier lives.

FIGHTING BURNOUT

Stress and burnout pose huge problems in our fast-paced world. We live in a culture in which we want everything now, and new terms like "FOMO" (fear of missing out) and "nomophobia" (no-mobile-phobia) show just how much we rely on constant connectivity.

Modern technology offers us huge advantages, but it also puts people under a lot of stress. In a given month, three quarters of Americans experience either physical or psychological symptoms related to stress, and half of Americans feel that their stress levels have increased over the past five years.[222] Statistics from the UK's Labour Force Survey show that 11.7 million working days are lost each year due to stress, anxiety and depression, with stress alone

---

[221] See: http://bit.ly/ibmwatsonpaultang

[222] See: http://bit.ly/stressstatistics

accounting for 45% of all lost working days due to ill health.[223]

Perhaps unsurprisingly, healthcare often comes towards the top of the list when people study the most stressful industries to work in. On top of that, medical doctors are the professionals with the highest suicide rate, with dentists coming in second. According to Mental Health Daily, "Examining all causes of death as a doctor, nearly 4% of all doctor deaths result from suicide. There are many factors that are believed to make doctors more likely to resort to suicide than average, including: long hours, demanding patients, malpractice lawsuits, continued education, medical school expenses and ease of access to medications."[224]

According to the annual Medscape Lifestyle Report, over half of physicians in many specialties – and primary care doctors and emergency physicians in particular – experience burnout. There's been a lot of talk of late that points the finger towards electronic health records as a major source of stress, and an article by Lloyd Minor for Quartz[225] puts its finger on the problem: "Together with the compressed time of office visits, EHRs conspire to turn medical practice into a regimented, one-size-fits-all endeavor, just when science and technology are giving us more ability than ever to treat our patients as the individuals they are. Today's search engines are better at helping doctors diagnose disease than our EHRs."

Minor knows what he's talking about. As the dean of Stanford University's School of Medicine, he's an influential figure in the field of healthcare and one of a growing number of people who identifies EHRs as a problem. One survey published in JAMA Internal Medicine seemed to back this up, going so far as to suggest that the technology was built to be a billing platform instead of as a tool to facilitate patient care.[226]

But not everyone agrees with Minor. Michael Blum MD, the

---

[223] See: http://bit.ly/lfsstats

[224] See: http://bit.ly/mhdoctors

[225] See: http://bit.ly/burnoutlloyd

[226] See: http://bit.ly/jamasurvey

chief medical information officer at the University of California San Francisco Medical Center, believes that physician burnout predated EHRs. He says that physicians started becoming more dissatisfied when increased regulation turned their notes into a billing document, adding, "Then [they] threw [EHRs] on top of that and made a bad situation horribly worse."[227]

Regardless of which side of the fence you stand on, we can all agree that something has to change. Better technologies need to be developed or more people hired to complete the paperwork to give physicians more time to see patients. Another (much better) alternative would be to get rid of the paperwork altogether, and voice-assisted medical transcription at the point of care will help to make this a reality.

There needs to be real action. Otherwise, doctors will have less and less time to actually examine and treat their patients. This will hurt us all and eventually the entire system will hit a breaking point. After all, no patient wants to hear, "The doctor will see your paperwork now." Do you?

RETHINKING EDUCATION

The University of Vermont has proposed a radical new approach to medical education with no lectures required – and as you might imagine, it's caused a huge amount of controversy.

Lecture-based learning is increasingly being replaced by team-based scenarios and problem-based learning which aim to put people to the test in real-world scenarios. After all, physicians won't be practicing in the lecture hall, and there's a limit to how much information the human brain can retain.

This approach actually makes sense, when you think about it. In

---

[227] See: http://bit.ly/physicianburnoutehrs

an article for The New England Journal of Medicine[228], Richard M. Schwartzstein MD and David H. Roberts MD argue: "Educators giving a traditional lecture with dozens of content-heavy PowerPoint slides may confuse what they teach with what students learn. The fact that a teacher has presented a piece of information does not mean that students have learned it. In fact, cognitive-load theory suggests that our brains are limited in the amount of information that they can process at a time. 60 slides in 45 minutes may seem like an efficient way to teach, but it is unlikely to be an effective way to learn."

The internet has already started to change the way that our brains work. In a 2011 experiment published in Science Magazine, college students remembered less information when they knew they could easily access it later on a computer. Meanwhile, neuroimaging of heavy internet users shows twice as much activity in the short term memory as it does in sporadic users. According to Academic Earth, "Our brain is learning to disregard information found online, and this connection becomes stronger every time we experience it. So the more we use Google, the less likely we are to retain what we see."[229]

It's certainly true that we're using information in new and interesting ways. The old style of teaching by rote is no longer relevant when the same information is available at the tips of our fingers. Instead of memorizing a list of every president of the United States, it's much more efficient to spend the same time teaching kids how to research information and how to decide whether the information is trustworthy or not. On top of that, people learn in different ways. Some people prefer learning by rote in a lecture hall, but others learn best by reading textbooks or through hands on experience.

Of course, hands on experience has always been difficult to come by in the healthcare field. It's a real risk to allow an unqualified

---

[228] See: http://bit.ly/newenglandjournalarticle

[229] See: http://bit.ly/academiceartharticle

trainee to perform a serious operation, but those risks disappear if they use new technologies like virtual reality. And that means by the time that they carry out their first real-world operation, they'll be chock full of experience that they've picked up from a virtual reality setting.

THE DOCTOR-PATIENT RELATIONSHIP

While I was working on this book, I spotted an article by Meghan O'Rourke in The Atlantic in which she talks about a slew of books by disillusioned physicians which reveal "a corrosive doctor-patient relationship at the heart of our healthcare crisis."[230] Perhaps this is one of those books, too.

O'Rourke describes her experience as a sufferer of a chronic disease and explains that when she was in hospital, she "always felt like Alice at the Mad Hatter's tea party." She says, "I had woken up in a world that seemed utterly logical to its inhabitants but quite mad to me."

It's a fantastic article that I recommend checking out if you've got the time, but the crux of it is that our modern medical system is "technologically proficient but emotionally deficient." And as O'Rourke points out, "This absence [of emotion] matters, because how patients feel about their medical interactions really does influence the efficacy of the care they receive, and doctors' emotions about their work in turn influence the quality of the care they provide."

He cites the words of Sandeep Jauhar, the author of *Doctored: The Disillusionment of an American Physician*[231], who believes that the medical profession is going through a midlife crisis. He says today's physicians don't see themselves as "pillars of [the] community" but

---

[230] See: http://bit.ly/doctorstellall

[231] See: http://bit.ly/doctoredjauhar

as "technicians on an assembly line" or "pawn[s] in a money-making game for hospital administrators."

The problem is that healthcare professionals just don't have the time they need to give their patients their full attention. EHRs are part of this, and so are networks such as large physician groups that force them to see a certain number of patients per day if they want to be well-paid for the work they do. No wonder we're so used to five-minute appointments.

O'Rourke points out that today's doctors and hospital workers are spending between 12% and 17% of their day with their patients, with physicians in non-hospital practices spending "ten times as many hours on nonclinical administrative duties" as the same staff across the border in Canada.

If the curricula of medical schools were based on what I experienced as a clinician, the bulk of the content would be on how to do the paperwork. Medical school admissions essays would be on "why I want to fill out paperwork". "Introduction to filling out forms" and "advanced form-filling" would be mandatory courses. The acronym MD would stand for "Mastered Documentation". This might sound like an exaggeration, but it isn't. A recent study that was published in the Annals of Internal Medicine found that for every hour physicians spent providing direct clinical face time to patients, nearly two more hours were spent on EHRs and desk work.[232]

What's even more troubling is that even when doctors are in examination rooms with their patients, they're spending only 52.9% of the time talking to or examining patients and spending 37% doing – you guessed it – paperwork. Outside office hours, physicians spend another 1-2 hours of personal time every night doing – drumroll please – additional paperwork.[233]

Is this really empowering doctors to practice at the top of their field? It's like if the Chicago Bulls told Michael Jordon to spend the

---

[232] See: http://bit.ly/doctorswastingtime

[233] See: http://bit.ly/physiciantimebreakdown

majority of his time manning the ticket windows and phone lines. Isn't it also wasting the time of patients, who came for the doctor's medical expertise and not his ability to do the paperwork?

According to O'Rourke in his article for The Atlantic, "The alarming part is how fast doctors' empathy wanes. Studies show that it plunges in the third year of medical school. That's exactly when initially eager and idealistic students start seeing patients on rotation." Doctors are overworked and underpaid, and the system gives us no time for so-called "slow medicine" in which physicians take their time to treat their patients. Critics call this a form of indulgence, but the truth is that it might actually make us more efficient. "More accurate diagnoses and effective low-tech treatments help the system save money," O'Rourke argues, "and result in fewer malpractice suits."

But what I particularly like about this article is the author's argument for putting emotion back into healthcare. O'Rourke says, "In the course of our lives, most of us will urgently need care, sometimes when we least expect it. Currently, we must seek it in a system that excels at stripping our medical shepherds of their humanity, leaving them shells of the doctors (and people) they want to be, and us alone in the sterile rooms they manage. What makes our predicament so puzzling and what may offer hope is that nearly all of us want a different outcome. I used to think that change was necessary for the patient's sake. Now I see that it's necessary for the doctor's sake, too."

And if you're not convinced of the importance of a strong doctor-patient relationship, take it from Hippocrates who said, "Some patients recover their health simply through their contentment with the goodness of the physician."[234]

---

[234] See: http://bit.ly/hippocratesquote

THE FUTURE OF JOBS

Will robots replace human workers? Some people seem to think so.

Robots have already replaced millions of manufacturing workers, and as automation continues to become more common, even doctors are starting to worry that their jobs might be at risk. This is backed up by Professor Ed Hess of the University of Virginia's Darden School of Business, who says that nearly half of American jobs could be automated, "including retail store clerks, doctors who scan X-rays for disease, administrative workers, legal staffers and middle managers."[235]

Fortunately, automation doesn't take jobs away from humans but instead redistributes them. This is the case in the banking industry, where the number of human tellers actually rose after the introduction of the ATM. The machines automated tasks like simple cash deposits and withdrawal, freeing up humans to carry out customer service and sales transactions.

One report by Redwood Software and the Center for Ergonomics and Business Research (CEBR) found that increasing investment and falling prices will help the robotics industry to grow. David Whitaker, an economist at the CEBR, said that while robots will take over mundane tasks, higher productivity could actually increase wages for human workers. Nevertheless, he also warns that people who want to stay in employment must focus on honing skills that robots can't handle – and that's exactly what physicians will need to do if they want to stay relevant in the future of healthcare.

The US Treasury Secretary Steven Mnuchin went on record to say it will be "50 or 100 years" before artificial intelligence takes over American jobs, adding, "I think we're so far away from that, [it's] not even on my radar."

I disagree. It's already happening.

---

[235] See: http://bit.ly/robotsmanufacturingworkers

THREE INSPIRATIONS FOR PHYSICIANS

I like to think that I'm able to approach the healthcare industry from a variety of angles, but it's also true that I have a particular affinity with practicing physicians. I read literally dozens of books, scores of studies and hundreds of articles while researching this book, but there are three books that stand out in particular – and especially for physicians.

So if you're a physician and you want to learn more about how the future of healthcare will affect your job and your life, you should read them. Meet Topol, Wachter and Gawande:

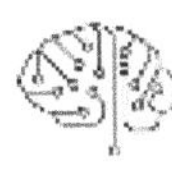

ERIC TOPOL MD: Topol is the author of *The Patient Will See You Now*[236], which largely focusses on how Moore's Law will boost healthcare delivery to the point at which digital medicine is democratized throughout the world – hence the book's tagline, "the future of medicine is in your hands."

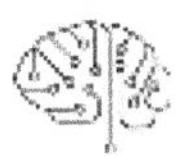

ROBERT WACHTER MD: Wachter is the author of *The Digital Doctor*[237], which is all about "hope, hype and harm at the dawn of medicine's computer age." It aims to make sense of the huge amounts of data that are created and, to quote Health Populi, "the cultural challenges physicians [face due to] cyberchrondriacal patients who over-Google symptoms and under-analyze search results."[238]

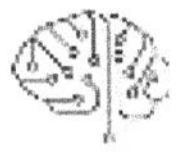

ATUL GAWANDE: As the author of *Being Mortal*[239], Gawande chooses to focus on the impact that our longer

---

236 See: http://bit.ly/patientwillseeyounow

237 See: http://bit.ly/digitaldoctor

238 See: http://bit.ly/healthpopuliarticle

239 See: http://bit.ly/beingmortalgawande

lives have on our overall well-being. When the author asked Dr. Felix Silverstone, a veteran geriatrician from New York, whether there was any reproducible pathway to ageing, Silverstone replied: "No, we just fall apart." He argues that the physician's job isn't to keep people alive but to ensure their overall wellbeing. Quality of life is more important than longevity. After all, without quality of life, what's the point of living?

WILL AI REPLACE PHYSICIANS?

The short answer to this is no.

It's analogous to the rise of electricity. Electricity didn't take people's jobs, but it did change the way that they carried them out. It didn't even replace steam, because steam turbines still generate most of the electricity that we use today.

It's AI's job to take increasingly complex tasks and to solve them at a rate that's faster than any human being could ever manage.

Billionaire engineer and businessman Vinod Khosla believes that AI will take away 80% of physicians' work, but he argues that it's different to simply taking their jobs away. Instead, AI will take on 80% of their existing workload so that they can spend that time focusing on "the human aspects of medical practice such as empathy and ethical choices." They'll be able to deliver more value to their patients by counselling them on weight management, opioid use, mental health issues, surgical questions, complications and a whole host of other things that patients both want and need.

Nevertheless, there are some interesting possibilities when we look at the rise of chat bots, which are now being used for everything from marketing and customer service to medical diagnosis.

One study carried out a direct comparison between human

physicians and 23 of the most commonly used symptom checkers to test their diagnostic accuracy.[240] The study found that the bots were "clearly inferior", scoring a 51% success rate compared to 84% for humans. That might not sound promising, but there are several factors to consider. The fact that the symptom-checkers were popular doesn't necessarily mean that they're using state of the art technology, and on top of that, humans were still wrong in 16% of cases. Humans aren't likely to get much better. AI, on the other hand, is constantly improving thanks to the constant flow of big data and the use of machine learning.

And AI might have one last ace up its sleeve. One study of the virtual therapist Ellie found that people opened up to her more when they were told she was an AI than they did when they believed she was human.[241]

WILL AI MAKE PHYSICIANS BETTER?

Yes. That's why this book is subtitled 'humans and machines partnering for better outcomes'. The truth is that both humans and machines work well in isolation, but that's nothing compared to what they could achieve together. It's a symbiotic relationship and a case of the whole being greater than the sum of its parts.

This brings me on to an article by Dr. Bertalan Meskó, The Medical Futurist.[242] He explains that "artificial intelligence will redesign complete healthcare systems in the near future, but it will also impact the life of the 'average doctor' positively." He goes on to share this list of ten of the ways in which AI could help physicians to be better at the jobs they do:

---

[240] See: http://bit.ly/aiversusphysicians

[241] See: http://bit.ly/ellievirtualtherapist

[242] See: http://bit.ly/aimedicalfuturist

1. **Eradicate waiting time**
2. **Prioritize emails**
3. **Find information**
4. **Keep doctors up-to-date**
5. **Work when doctors aren't working**
6. **Help to rationalize hard decisions**
7. **Help patients with urgent needs to get in touch**
8. **Help doctors to improve over time**
9. **Help doctors to collaborate more**
10. **Do the admin work**

As an industry, we're quick to talk about the high level stuff, the super exciting uses of technology that can save lives and discover new drugs. But the truth is that it's the more 'minor' applications that will make the real difference, making doctors more efficient without them even realizing it. That's good news for physicians, but it's good news for patients too.

Another intriguing technology is Enlitic, which uses deep learning to allow doctors to make faster and more accurate decisions. With over 25,000 petabytes of patient data expected by 2020, a solution like Enlitic is vital if we want to make the most of it. Enlitic's developers claim to "[use] deep learning to distill actionable insights from billions of clinical cases" and to be able to "interpret a medical image in milliseconds – up to 10,000 times faster than the average radiologist." It even allows you to carry out retrospective analysis to identify false positives, both in the hospital setting and during clinical trials and drug development.[243]

Perhaps Abhinav Shashank, cofounder and CEO of InnovAccer, put it best when he said: "Healthcare is a people business, and people make mistakes – though mistakes are precisely what healthcare technology is designed to minimize."[244] AI won't take doctors' jobs – it'll just make sure that they're doing them properly.

There's a famous Latin phrase from the Roman treatise *Satires of Juvenal* which goes "quis custodiet ipsos custodes?" and means "who will watch the watchers?" Perhaps it would be more apt if the phrase was "quis custodiet ipsos doctorum?"

THE TROUBLE WITH EHRS

Electronic health records (EHRs) were originally created to help doctors, but now they're a hindrance at best, turning talented physicians into little more than clerks whose job it is to feed the machine with more data. A cynic would say that doctors are paying for the privilege of giving patient data to the EHR company, which the company can then sell to healthcare companies for a profit. There are all sorts of apocryphal stories out there about how EHRs do more harm than good, such as one I heard recently about a doctor saying that no matter what you do, you should never admit that a patient has diabetes because if you do, you'll have to spend the next twenty minutes clicking through screens instead of treating the patient.

On top of that, many practicing doctors are forced to follow "embedded clinical decision supports", which is a fancy way of saying that computer software is telling them what to do. Unfortunately, technology itself doesn't solve the problem – and in this case, it makes it worse by forcing doctors to follow a set of instructions no matter what the patient actually wants.

---

[243] See: http://bit.ly/enlitic

[244] See: http://bit.ly/peoplebusinessshashank

Doctors aren't even involved in the design of EHRs, which means that the people building the systems actually have no idea how to make doctors' lives easier. You can't design something well if you don't use it yourself. Dr. Bruce Y. Lee, associate professor of international health at the Johns Hopkins Bloomberg School of Public Health, wrote in a recent article that "[it would be equivalent to Lady Gaga designing NFL football uniforms. Platform shoes and dresses made out of meat just wouldn't work on the Dallas Cowboys and Los Angeles Rams."[245]

In an article for Investor's Business Daily, Betsy McCaughey said, "Jeffrey Moses, an interventional cardiologist at Columbia-Presbyterian, complains that the computer prompts are 'taking the doctor's eyes off the patient.' Another cardiologist compares it to being 'demoted to an airline booking agent.' Dr. Lloyd Minor, dean of Stanford University Medical School, says 'there is nothing more frustrating to a patient than talking to their doctor, wanting advice, and that provider is typing away and looking at a computer screen.'"[246]

A byproduct of these systems is that every time a patient comes in, the doctor is forced to recite the same set of questions about sleeping, breathing, digestion and other systems, regardless of whether the patient answered the same questions at their last appointment. This applies even if it's a follow-up meeting and takes time that could be better spent elsewhere. Dr. Jeffrey Borer, a heart valve specialist from New York, calls it "cookie cutter medicine" because if the patient's needs don't meet the pre-set protocols, they're out of luck. It's about as far away from personalized healthcare as it's possible to get.

Jeffrey Moses, the interventional cardiologist interviewed by Betsy McCaughey, summarized the system by describing his disdain at having to fill out irrelevant information despite "[getting] no points for identifying a disease that went undiagnosed for five

---

[245] See: http://bit.ly/ladygaganflquote

[246] See: http://bit.ly/betsymccaughey

years."

Howard Green MD has similar concerns. He explained: "Their advertisements and marketing state that their EMR is, 'Transforming how healthcare information is created, consumed and utilized to increase efficiency and improve outcomes.' Today, after approximately eight months of using EMRs in our practice, I and the dozen doctors in our practice can definitively report that our EMR has a negative value on our patients' outcomes and our practice. There is not one aspect of our electronic medical records system which has made our practice more efficient, productive, and safer, improved our clinical outcomes, saved our patients or ourselves expenses or time or facilitated our compliance with federal regulations."[247]

CLINICAL DECISION SUPPORT TOOLS

Let me be clear here: I'm not saying that EMRs are totally useless. The problem is that the current system is crippled and has no hope of ever living up to its full potential. For EMRs to become useful to all stakeholders, they need to be dramatically rethought and completely interoperable. They need to be supplemented by artificial intelligence so that they don't take time away from busy physicians. They need to add value to patients first and doctors second, with the EMR companies left as just an afterthought.

One of the most exciting possibilities when it comes to EHRs is the use of clinical decision support (CDS) tools. Loosely speaking, these are designed to offer providers, patients and other stakeholders with person-specific information at the right time so that it enhances the approach to health and healthcare.

A great example of this is RxRevu, the world's most advanced

---

[247] See: http://bit.ly/emrbuyerbeware

prescription decision support platform.[248] It's designed to help prescribers to determine what medication their patient needs based on factors like whether there's a gap in care or whether a more cost-effective alternative is available. It even offers real-time insights and predictive analytics to make prescribing decisions easier than ever before.

Tools like these will be able to tap into EHR data to provide bespoke advice for every patient, and they'll also be able to gather insights on the patient population as a whole. This will increase quality of care for all and lead to a future in which EHRs are no longer a nuisance that providers have to battle with. Instead, they'll be every physician's best friend and most loyal ally.

THE DOCTORS OFFICE OF THE FUTURE

San Francisco healthcare startup Forward, which is the brainchild of veteran Google and Uber staffers, has raised $100 million in funding and is showing off its vision of the future of healthcare by hitting the road with the doctor's office of the future.

The company's concierge clinic comes with tablet computers, biometric body scanners, genetics tests, interactive screens, body-monitoring wearables and more, and they took the concept to the general public by putting it inside an 18-foot trailer and taking it on tour.

What's interesting about Forward is their route to monetization. They charge customers $149 a month in exchange for access to their clinics and the ability to reach out to doctors and nurses through emails and text messages. This new approach to healthcare isn't covered by health insurance, but it does show how things could work when we start to put value-based healthcare first.

In their coverage of the road trip, Buzzfeed's Stephanie M. Lee

---

[248] See: http://bit.ly/rexrevu

explained, "Forward isn't comprehensive: It doesn't cover services like hospitalizations, surgeries, or specialist care. Instead, the company says it focuses on encouraging people to proactively take care of their health before problems and conditions develop."[249]

That sounds like the future of healthcare to me.

[249] See: http://bit.ly/forwardarticlebuzzfeed

CHAPTER ELEVEN: THE FUTURE FOR HOSPITALS

"WE STILL HAVE WORK AHEAD TO GET THESE ALGORITHMS INTO THE HEALTHCARE SYSTEM'S WORKFLOW. BUT I THINK HEALTHCARE 10 YEARS FROM NOW WILL USE A LOT MORE AI AND WILL LOOK VERY DIFFERENT THAN IT DOES TODAY."

- ANDREW NG, VP & CHIEF SCIENTIST OF BAIDU

IT'S NOT JUST PATIENTS, physicians and pharmaceutical companies that will face disruption as they enter the future. Hospitals are also likely to face a more pleasant type of disruption which could help to free up their funding. As smartphones and commercial wearables become more and more powerful, they'll be able to take on more of the roles that were previously the domain of specialized machines.

Just take a look at Schwetak Patel, Professor of Computer Science and Engineering at the University of Washington. He used the internet of things to combat chronic obstructive pulmonary disease – also known as COPD, chronic bronchitis and emphysema. Diagnosing the disease requires the use of a spirometer, which measures the flow of air in and out of the lungs. Unfortunately, these cost thousands of dollars and are not always available, so Patel spearheaded the creation of an algorithm that can measure lung health by analyzing the sound of someone blowing on their smartphone's microphone[250] – with a 95% success rate, even on landlines. This is great news for all parties – including patients, hospitals, physicians and insurers – because it makes it easier to diagnose COPD and start treatment. Anything that makes existing technologies available via a smartphone will help to cut down operational and equipment costs for cash-strapped hospitals, enabling them to focus their budget elsewhere.

WAYFINDING

In his book, *Making Sense of IoT: How the Internet of Things Became Humanity's Nervous System*, Kevin Ashton talked about the issue of wayfinding, a surprisingly common debate amongst hospital workers and the healthcare industry. He explained that when people get lost in a hospital, it causes them stress and costs them money, with one study at Atlanta's Emory University Hospital finding that wayfinding problems cost $400,000 per year, or just over $800 per bed. Most of this expense comes from staff being interrupted to provide directions.

The problem is exacerbated by the fact that hospitals are often housed in old buildings, which can be difficult to navigate. That's why hospitals of the future will need to follow in the footsteps of Boston's Children's Hospital, which developed a GPS-based smartphone app to help direct patients to their destinations. There's no reason why they couldn't have taken this a step further, installing voice assistants throughout the hospital which could respond to voice queries and provide accurate directions. Meanwhile, the internet of things could provide real-time information on the hospital's facilities, so that if an elevator is out of order or a corridor has been closed, visitors can be re-routed.

---

[250] See: http://bit.ly/iotinboston

This comes back to a concept that we've seen elsewhere – that of hospitals needing to act more holistically, making the patient's experience as stress-free as possible. Providing voice assistants and better signposting will be a big part of this, but there's much, much more for hospitals to do if they want to lead the charge into the value-based future.

NEW TECHNOLOGIES

One of the major challenges for hospitals will be the installation and implementation of new systems. They'll need to provide training for staff members and create a culture in which adapting to new technologies is the norm.

Technology is already changing the way that healthcare works. Just look at M*Modal and its NoteReader implementation, which applies machine learning and clinical reading to electronic health records (EHRs).[251] What's particularly interesting is that it works in the background at first, improving its understanding of operations without needing human input. Once it's up and running, it can offer feedback on health records – either in real-time or before they're saved – to help doctors to fight 'alert fatigue'. Essentially, it'll help to improve documentation and reduce the amount of time it takes to keep records, freeing up more time for doctors to interact with their patients.

Meanwhile, the increase in patient technology and the widespread adoption of new technologies and artificial intelligence have the potential to usher in cost savings. A team of researchers from Google's DeepMind and Stanford University have used deep learning to accurately interpret medical scans on a par with human doctors. According to biomedical informatics specialists Andrew Beam and Isaac Kohane, who discussed AI in a paper in the Journal

[251] See: http://bit.ly/anepicpartnership

of the American Medical Association, computers can now read 260 million medical scans in a single day at a cost of just $1,000.[252]

Then there's Ambra Health, a cloud-based image platform which helps with kidney transplants by saving coordinators 1,400 hours a year by allowing them to upload images to the cloud.[253] These images can be accessed in near real-time at transplant centers from all over the country, which is why the platform is used by the National Kidney Registry. Providers can transfer kidney donor and recipient information between team members in a matter of minutes, whereas the old system took at least ten days.

Joe Sinacore, the director of education and development at the National Kidney Registry, explains, "The way we did this for the first eight years was, unfortunately, transplant coordinators had to go to the radiology group, get a copy of a CD with all these images on them and then [use expedited shipping services to send it] to the recipient centers."

This process could cause vital delays and even stop surgery from going ahead altogether, so it's easy to see the benefits that such a relatively simple cloud-based system could have to offer. And, as we discussed back in chapter seven when we took a look at The Secret Rules of Modern Living, the approach can even help to create chains of donor/recipient pairs to overcome the lack of a match amongst close friends and family members who are willing to donate a kidney. The National Kidney Registry has created more than 400 of these chains using Ambra Health and has transplanted over 2,200 kidneys as a result of it. It's no wonder that the NKR awarded Ambra Health with its prestigious Terasaki award.[254]

Research and innovation in the field of kidney transplants kept on cropping up while I was writing this book, more so than any other type of transplant. Carnegie Mellon also developed an algorithm to match living kidney donors with medically compatible

---

[252] See: http://bit.ly/thepatientwillseeyounow

[253] See: http://bit.ly/ambrahealth

[254] See: http://bit.ly/ambrawin

transplant candidates, a system which is similar to the one used by the NHS.[255] Meanwhile, Wired published an article called *The Science Behind a Crazy 6-Way Kidney Exchange* which talked about operations resulting from another similar system called MatchGrid. It was described as "a surgery worthy of the most convoluted Grey's Anatomy plot" and "the West Coast's largest paired transplant ever."[256]

PREDICTIVE ANALYTICS

Predictive analytics is a rapidly growing new field in which historical data can be used to identify potential trends in the future. It could change the way that hospitals operate if used correctly, and it could also bring about better health outcomes by ensuring that any shortfalls in resources are identified beforehand.

An example of this in action comes to us via UnityPoint Health, an Iowa-based company which used predictive analytics to reduce readmissions by 40% over a period of three years at a pilot hospital. They did this by integrating predictive analytics into care teams' daily workflows based on the idea that insights are only useful if you act upon them. By putting the analytics front and center, they made sure that employees were able to make tangible, real world decisions based on the analytical information they were presented with.

There's a fantastic article by Mike Miliard in Healthcare IT News which dives into exactly how the system worked, but I'm not going to go into too much detail here. Ben Cleveland, a data scientist at UnityPoint Health, summarized it best when he said: "What we found is that some patients were at much greater risk of coming back early on after their stay and then others tended to be more at risk later on in that 30-day timeline. Maybe their problems would

---

[255] See: http://bit.ly/carnegiemellonorgans

[256] See: http://bit.ly/wiredkidneyscience

compound over time or they would miss their follow-up appointments or they wouldn't follow directions for medication. So we developed a risk heat map over that 30-day timeline that visually depicts a patient's risk very quickly. You don't have to be a data scientist to interpret the output you're looking at – you can look at it quickly. For every patient we have they have their own individualized heat map that our care teams are working off of."[257]

When you know that a patient is at risk you can stage an intervention. This applies whether they forget to take their medication or whether they have a high probability of not turning up to a follow-up. By intercepting problems before they happen, we can practice a more holistic form of medicine which focuses on prevention and interception instead of mitigating the symptoms with endless amounts of medication when the condition has irreversibly deteriorated.

William J. Mayo, one of the founders of the Mayo Clinic, famously said that "the aim of medicine is to prevent disease and prolong life" and that "the ideal of medicine is to eliminate the need of a physician."[258] I'm not convinced that there'll ever be a future in which physicians aren't needed, but we can eliminate the need for patients to see doctors about preventable issues and minor illnesses. This will free up their time so that they can spend it with the patients who really need it.

AI-focused healthcare startups that focus on predictive and preventative medicine are on the rise. You just need to take a look at PubMed to see it in action. Out of 218 AI healthcare startups, 54 were involved in predictive medicine, and the upward trend will only continue to grow exponentially as new research pushes the industry towards the future. Prediction and prevention are well-known concepts for healthcare professionals. Now they're being brought into the 21st century thanks to the rise of machine learning and predictive analytics.

---

[257] See: http://bit.ly/unitypointsystem

[258] See: http://bit.ly/mayoclinicquote

USING ALEXA

Alexa is already useful around the house, but it could usher in a whole new era of human/AI cooperation if Amazon continues its push into the healthcare industry.

Hospitals are already using Alexa in a range of circumstances. Some surgeons use an Amazon Echo during operations to carry out basic searches or to control music during long hours in surgery. Many patients are surprised by the fact that surgeons listen to music while they're in the theater, but there's evidence to suggest that listening to music boosts productivity and that music is played just as often in the operating room as it is in corporate offices.[259] According to data from Spotify, rock (49%) is the most popular genre for surgeons, followed by pop (48%), classical (43%), jazz (24%) and R&B (21%). 89% of doctors put on their own playlists instead of listening to albums and top tracks in surgery include Cocaine by Eric Clapton, Beautiful Day by U2, Piece of My Heart by Janis Joplin and Rock You Like a Hurricane by The Scorpions. Transplant surgeon Dr. Alan I. Benvenisty from New York City says, "People's lives are in my hands and listening to rock puts me in a comfortable place so my full attention is on my patients."[260] During my own time spent in plastic surgery, Bob Marley and Fela Kuti were among my favorites.

So it's easy to see how Alexa could help out in a hospital setting, if only as a DJ for doctors so that nurses don't have to spend their time working the 'turntable'. Alexa could also one day be used for everything from providing directions to patients to documentation and transcription, although it'll only be able to do this if guidelines and regulations change to make innovation easier for hospitals and healthcare providers.

Who knows? Perhaps Alexa could even save lives. As CNBC explains, "Physicians at Massachusetts General Hospital are

---

[259] See: http://bit.ly/musicproductivity

[260] See: http://bit.ly/spotifydoctors

researching how text-to-speech technology can be useful in helping surgeons comply with surgical safety checklists in the operating room. Paul Uppot, a radiologist at the hospital, told CNBC that in one case a patient was listening in to the safety checklist via a voice application right before going under. The patient had an allergy to latex, a fact that was missing from the medical record, but was addressed thanks to the checklist. That might have averted disaster as the surgeons had intended to use latex gloves."[261]

CYBERSECURITY

One of the biggest news stories to hit the healthcare industry over the last couple of years has been the high-profile WannaCry cyberattack that hit the NHS – as well as many others. The Telegraph explained that "more than 300,000 computers were infected while the countries most affected by WannaCry were Russia, Taiwan, Ukraine and India." Mikko Hypponen, chief research officer at Helsinki's F-Secure, said the attack was "the biggest ransomware outbreak in history."[262]

The fact that the NHS was hit by the attack goes to show that hospitals need to invest heavily in IT security. No private company the size of the NHS would have allowed itself to be so vulnerable, but it's hard to lay the blame at the organization's metaphorical feet. They're under enough pressure as it is thanks to funding cuts and staff shortages. They simply don't have the money, the resources or the knowhow to deploy secure infrastructure, and hackers are quick to take advantage of that. The only surprise is that it took this long for the NHS to be affected.

The sad truth is that if hospitals fail to protect their infrastructure, it'll continue to be compromised in the years to come.

---

[261] See: http://bit.ly/alexasavinglives

[262] See: http://bit.ly/WannaCryattack

This has a knock on effect when it comes to earning the trust of the general public. Why would anybody trust their healthcare practitioners with their data if they're unable to store it securely?

A lack of budget seems to be one of the main reasons why healthcare facilities have historically failed to invest in cybersecurity. Unfortunately, they're just going to have to find the money from somewhere. I'm not saying the change will be easy, but I am saying that hospitals are going to have to move with the times or get left behind. There's room for everyone in the future of healthcare, but you have to earn your spot at the table.

CHAPTER TWELVE: THE FUTURE FOR INSURERS

"A CENTURY FROM NOW, NOBODY WILL MUCH CARE ABOUT HOW LONG IT TOOK, ONLY WHAT HAPPENED NEXT."

- GARY MARCUS, PROFESSOR OF COGNITIVE SCIENCE AT NYU AND AUTHOR OF GUITAR ZERO

ACORDING TO ANIL JAIN, chief medical officer at IBM Watson, "about one third of every dollar spent on health is probably unnecessary."[263] And unfortunately for insurers, somebody needs to foot the bill.

That's why insurers in the future of healthcare will need to follow in the footsteps of Aetna, which is transforming from an insurance company to a health company by helping members to "achieve their life goals by achieving a state of health." The company's president of transformative markets was featured on the Oliver Wyman Health podcast[264] to talk about the changing state of health insurance, and it's worth giving it a listen if you're curious about why modern insurers are rethinking their position.

One interesting opportunity for the industry is the growing rise of automation. This could have a huge impact for insurance

---

[263] See: http://bit.ly/ibmwatsonpaultang

[264] See: http://bit.ly/aetnatransformation

companies because much of their work could be automated, particularly when it comes to logging and processing information. By automating the work, they could reduce their operating costs and pass the savings on to the patients. Alternatively, it can free people up to carry out activities that are more beneficial for patients and to focus on providing a valuable service, instead of just filling out the paperwork.

Meanwhile, insurers are facing challenges from new payment plans and innovators in the healthcare space, as well as new startups that promise to democratize insurance and to create a system that works for everyone.

Let's face it: insurance companies aren't popular. They're seen by most as a necessary evil, while many more will flat out tell you that insurance companies were "created by the devil". But I'm not here to bash insurance companies. Many of them are doing a fine job considering the system in which they find themselves. They're starting to catch on to what's happening and putting things in place to address it. Here's what that looks like in practice.

CUTTING OUT THE MIDDLEMAN

One of the problems with the current system is that the business model behind it is all about making money, money, money for the middlemen to the detriment of the patient. Some insurance companies are striking deals with pharma companies in which they agree to favor brand name products over the cheaper generic options. This means that insurers pay less but the patients pay more, even when there's a cheaper option that would save the system more money overall. Perhaps most damning of all is the fact that this is happening behind closed doors and patients are being given no details about what's actually happening.

Take the case of Adderall XR, which is used to treat attention-deficit hyperactivity disorder (ADHD). As Charles Ornstein of ProPublica and Katie Thomas of the New York Times explain,

"Shire, the maker of Adderall XR, and some other brand-name drug manufacturers are no longer content to allow sales of their products to plummet when generic competitors arrive on the market. Instead, they're negotiating deals with insurers and pharmacy benefit managers to give priority to their versions. In the mid-2000s, Shire sued generic drug companies to stop them from bringing cheaper copies [to market], alleging patent infringement. Then it made deals with two makers of generic drugs to sell authorized copies of its drug, a tactic in which the branded manufacturer supplies its products in exchange for a share of royalties."[265]

Both the cause and the effect of this behavior is pretty clear. Pharmaceutical companies are paying insurers because it helps them to make more money, and insurers are accepting these payoffs for the same reason. As a result of that, we're spending more money than we need to on prescription medicine, both as a nation and as individual patients. No wonder Americans spend more per person on healthcare than anybody else.

DATA-CRUNCHING: REDUCING THE SIZE OF THE HAYSTACK

It's not just providers and patients that benefit from the use of artificial intelligence and big data. These technologies are already being used as a form of preventative healthcare by applying predictive analytics – which I introduced in the previous chapter – to identify people who are at risk of certain conditions, such as diabetes. Identifying these conditions early can drastically improve patient prognosis and potentially offer an opportunity to head diseases off before they become a problem. It could also disrupt the field of medical insurance.

As Brian Gormley explains in an article for the Wall Street

---

[265] See: http://bit.ly/genericdrug

Journal, "Startups such as Ayasdi Inc., GNS Healthcare Inc. and Prognos are providing insurers with data-analysis capabilities to help them forecast emerging illnesses and deploy resources. This can help them to better manage chronic diseases and improve the overall health of their members." [266] The idea is essentially that by stopping patients from developing an illness, it saves the insurers from expensive payouts down the line.

Crunching the numbers can help insurers in other ways, too. One example comes via UnitedHealth, which has been using Ayasdi's AI capabilities to hunt down fraudsters since 2014. The goal is to stop fraudulent claims from being paid in the first place instead of waiting until after the fact and then trying to recoup the payments. Then there's the use of AI and predictive analytics to determine the risks of infectious disease outbreaks, such as the threat from Zika and Ebola.

Richard Popiel, chief medical officer of Cambia Health Solutions, summarized the effect of these new technologies by saying, "In the past, we looked at a big haystack for a needle. Thanks to new analytical models, we've [dramatically reduced] the size of the haystack."

DISRUPTIVE TECHNOLOGY

As we've already seen from the rest of this chapter – not to mention the rest of the book – there's a huge amount of potential for new technologies to come along and disrupt the insurance industry. For instance, it could be used to collect real world data about patients to make premiums more accurate and improve the standard of the treatments that physicians are able to offer. At the same time, it could reduce costs for all involved by reducing wastage and offering more accurate decisions.

---

[266] See: http://bit.ly/wsjarticlegormley

AETNA is using this new technology to look at patients' results on a series of metabolic syndrome diagnostic tests, assessing risk factors and focusing treatment on the one or two things that will have the most impact (statistically speaking) on improving the patient's health.

Meanwhile, as I was coming to the end of the first draft of this book, Apple's COO Jeff Williams unveiled the Apple Watch Series 3, which Mashable described as "more health-focused than ever."[267] The previous model included a heart monitor, but the new one comes with a whole heap of new and improved features that make it a true medical device. As well as simply monitoring your heartbeat, it'll monitor your resting heart rate and recovery rate after exercise, and it'll even alert you if your resting heart rate is elevated while you're not doing any exercise. This could be a vital early warning sign for people with known and unknown heart conditions.

It's interesting to note that Apple, which is arguably one of the more insular of the modern day tech giants, did something that the healthcare industry could not. It listened to people. "[The new feature is] inspired by a lot of the letters we receive from customers who notice an unusually high heart rate when they wouldn't expect one," Williams explained. Mashable explains that "a rise in heart rate could be caused by such things as pregnancy, caffeine or anxiety, in addition to more serious issues."

At the same time, it was also revealed that Apple has partnered with Stanford University to carry out the "Apple Heart Study", which will use data from the Watch to study heart arrhythmia. The Apple Watch has already been used to identify arrhythmia, and the study will allow any Apple Watch user to sign up and participate.

One final example of new technology is the Beam toothbrush. It's an interconnected device which comes for free when you take out an insurance plan through the company, and you can also buy it as a standalone device for $49. Fortune explained, "To a dental insurance company, giving connected toothbrushes to policyholders

---

[267] See: http://bit.ly/thenewapplewatch

makes all the sense in the world. Knowing that your policyholders brush their teeth on the regular means they're less likely to develop cavities and other issues associated with high claims. The insurer might even be able to promote more brushing or even flossing using incentives from the app associated with the connected toothbrush."[268]

Beam even developed its own line of dental floss and toothpaste that's shipped to customers every quarter, as well as a nationwide network of 100,000 dentists. The goal is to provide preventative measures so that they can offer a leaner insurance plan. After all, if they can stop their customers from needing treatment then they can charge a lower premium and make a larger profit. It's a no brainer.

These new innovations have huge implications for patients, as it's ultimately they who will become the end users. But they're equally disruptive for insurers because these new data streams will provide them with a more accurate way to assess the health of their customers and to provide more accurate premiums and treatments as a consequence. It could save patients' lives, but it could save insurers' money, too. That's a win for everyone.

PROJECT BASELINE

One of the most promising developments over the last few years has been Google's Project Baseline, which has been deployed under the umbrella of Verily Life Sciences, Google's research organization. It's described as "a broad effort designed to develop a well-defined reference, or 'baseline', of good health, as well as a rich data platform that may be used to better understand the transition from health to disease and identify additional risk factors for diseases."

The company partnered with Duke University and Stanford Medicine to "collect, organize and analyze health data from

---

[268] See: http://bit.ly/beamtoothbrush

approximately 10,000 participants over the course of four years." Each of those participants was given a free tracking device to wear, and the data that they received was analyzed to see exactly what happens to a large group of people over a long period of time.

This is the future of healthcare. Imagine if when people were born, they were given a device that tracked their healthcare throughout their entire life. It could even be fitted with digital information such as their birth certificate and, as they get older, their driving license, their passport and their bank account. And it makes sense to keep all of this data together, especially if it's in an implant or a device like a smartphone that people already carry around. Plus, purchasing data could be used to inform healthcare – so if their doctor tells them to quit smoking and they buy a pack of cigarettes, he'll know.

Better still, if you could track your data from the day you're bon until the day you die, you'd really get to see what was happening. This would benefit you as a patient, but it would also help insurers. They'd be able to say, "Well, you walk a lot and get a lot of exercise. You have one or two minor health conditions but nothing too serious, so you're at a lower risk. We'll give you $50 off your premiums."

This brings us on to health scores, systems that are used by mobile health companies to track users' health over time. It's designed to form an overall rating of a user's health and to make the invisible and intangible aspect of 'health' something more measurable.

Imagine the impact that such a health score could have. It could help health and life insurance companies to determine the premium that a customer pays based on their health risk. People who exercise, eat well and generally take care of themselves will pay less for healthcare and insurance than people who don't. The data that we collect from mobile devices and other sources, combined with artificial intelligence and machine learning, will predict the risk of a patient developing diseases in the future. Meanwhile, physicians will know what their patients' lifestyles are like outside of the doctor's office or the hospital. Instead of them asking, "How have you been feeling since your last visit? Any improvements in physical activity?", they can simply receive a score and a physical activity report.

This is bringing real world evidence to the doctor's office, enabling the doctor and patient to have a real and objective conversation. It eliminates recall bias and leads to measures being put into place to improve patients' health proactively instead of waiting for diseases to get worse. That, of course, will decrease overall healthcare costs and allow insurers to run more profitably while still passing on savings to the end user.

THE RISE AND FALL OF TURNTABLE HEALTH

I should preface this by adding a disclosure here: Zubin Damania, also known as ZDoggMD, is the brains behind Turntable Health, and he's also a good friend of mine. He's also a good friend of Zappos.com's CEO Tony Hsieh, who asked him to launch a next-generation medical clinic in downtown Vegas as part of a $450 million revitalization project.

Turntable Health is interesting because it challenges the entire medical industry and its fee-for-service model. As Robert Pearl explains in an article for Forbes, "Inside a waiting room that resembled a sleek Silicon Valley startup, Turntable members passed the time by spinning records, playing Xbox and practicing yoga. As part of their membership, patients had access to an entire 'wellness ecosystem', complete with same-day visits, 24/7 doctor access (by email, phone or video), along with a dedicated health coach. Doctors at the clinic spend 45 minutes or more with their patients, quite unlike the 13 to 16-minute visits that have become standard in US doctors' offices."[269]

Turntable patients don't pay for each visit and procedure. Instead, those who aren't covered by their insurance simply pay a

---

[269] See: http://bit.ly/turntablehealth

flat fee of $80 a month. The idea is to focus on population health and disease prevention to improve patient wellness and lower costs. Pearl says, "By emphasizing prevention and doctor-patient relationships, Damania's practice achieved superior quality outcomes while providing rapid access to care and high patient satisfaction. But from an economic perspective, the clinic was a bust. Insurers shied away from member fees, insisting on more traditional reimbursements, which directly contradicted Damania's long-term health strategy. Turntable Health was forced to close its doors in January 2017, just three years after opening."

Some things are just ahead of their time, but in his typical style, Damania is sticking to his guns. He released a statement explaining, "We flatly refused to compromise when pressured by payers to offer fee-for-service options or to begin charging a co-pay. We firmly believe that healthcare is a relationship, not a transaction."

The problem that Damania came up against is that it takes time to see the results from investing in primary care and chronic-disease management, and insurance companies are simply impatient. They also face the potential threat of patients switching insurers before the initial company is able to recoup that long-term investment.

Still, the change that ZDogg is advocating will come about eventually. It has to. It's all about bringing real world evidence to the doctor's office, thereby enabling the doctor and patient to have a real and objective conversation. It eliminates recall bias and leads to measures being put into place to improve patients' health proactively instead of waiting for diseases to get worse. That, of course, will decrease overall healthcare costs and allow insurers to run more profitably while still passing on savings to the end user.

THE FUTURE OF INSURANCE

Let me tell you a story. This is a true story, and while it's sad, it's also an example of how our health and life insurance will work in the future.

It starts with would-be salon-goer Deshania Ferguson, who noticed a sign at her local salon which read: "Sorry, but if you are overweight, pedicures will be $45 due to service fees for pedicurists. Thank you!"[270] That worked out at over double their standard price, and it caused an understandable outrage when she shared the news on Twitter.

But as always, there are two sides to every story. In this instance, salon owner Son Nguyen explained that it's more difficult – and thus more time consuming – to give a pedicure to an overweight person, and that he's had two $2,000 pedicure chairs broken "as a result of overweight customers sitting in them."

It's not my place to say whether Nguyen is right or wrong, but it's a sign of the way things might be in the future. Overweight people are likely to face higher insurance costs because they're more likely to suffer from heart disease and other health problems. This shouldn't be seen as discrimination, though. It's just that everything about us will be used to identify the risks we face. Smokers will be charged more. So will people who don't get any exercise. It's really no different to how life insurance companies use your job and your lifestyle factors to determine the amount you'll need to pay for your premiums.

The focus of healthcare will be on prevention and living healthy. Those who don't live a healthy life will pay more for insurance. Technology, the internet of things and big data will make this happen through the use of predictive analytics.

MARK CUBAN'S NEW MODEL FOR HEALTHCARE

Perhaps the healthcare industry will find an unlikely savior in the form of billionaire Mark Cuban, who thinks he's found a model

---

[270] See: http://bit.ly/nailsalonexample

for fixing healthcare. His idea involves moving a little closer to Britain's NHS or Australia's Medicare, and he believes he could cut costs by over 50% by removing the role of private insurers.

As CNBC reports, the system involves "scrapping insurance companies from the US healthcare system and instead using federal funds to boost medical staff numbers and make care more widely accessible."[271] He shared his thoughts in a Twitter rant which started with a simple question: "Dear politicians. If every person in our country had health insurance, would we be any healthier?"

Cuban continued to explain that there's "no chance a system where you give [an insurance company dollars] then beg them to spend it among limited options is the way to optimize our healthcare." Perhaps he's right and perhaps he's wrong. We all have our own opinion, and that's what makes America the country that it is in the first place.

Personally, I think that despite all the talk about policies, single payer systems and the like, nothing changes the fact that our healthcare system – be it in Australia, the UK or the US – should be aimed at disease prevention. That's a guaranteed way to improve outcomes for every stakeholder.

---

[271] See: http://bit.ly/markcubannewmodel

CHAPTER THIRTEEN: THE FUTURE FOR GOVERNMENTS

"'DATA! DATA! DATA!' HE CRIED. 'I CAN'T MAKE BRICKS WITHOUT CLAY.'"

- SHERLOCK HOLMES

ONE OF THE BIG CHALLENGES that the government will face is the changing role of the Food and Drug Administration (FDA).

As of now, there isn't a clear regulatory framework for digital health. In the United States, the FDA is working alongside patients and the healthcare industry to bring the regulatory framework closer to the emerging technology frontier. There's a clear recognition of the potential for digital health to transform healthcare delivery. The FDA's Center for Devices and Radiological Health (CDRH), which oversees this area, says it puts patients at the forefront of its vision and is driven by timely patient access to high-quality, safe and effective medical technology. It holds the belief that digital technology is driving a revolution in healthcare, from mobile medical apps and fitness trackers to software that supports the clinical decisions that doctors make every single day.

With that being said, it's worth mentioning that one of the key challenges in the US (and other markets) is finding an appropriate business model for these tremendous new innovations. You see, healthcare has a unique problem. In other industries, such as retail and banking, technological innovation tends to increase the amount of choice, boosting overall performance while cutting costs. In the healthcare industry, while innovations still increase the amount of choice and demand, they also tend to increase prices. As a result, healthcare costs have been rising faster than inflation in the United States and around the world. To add insult to injury, most healthcare payments and reimbursements tend to be based on activity rather than on outcomes. That means that there's usually no incentive to focus on prevention because then they can't send you an invoice for providing the cure. This is a massive problem for healthcare which the government – and its regulatory bodies – could help to solve.

In June 2017, the FDA announced plans to increase its presence in the digital health space just six months after the US Congress passed the 21st Century Cures Act, authorizing over $6 billion in healthcare funding.[272]

The FDA is still trying to determine its place in the future of healthcare, and there are two different ways that they could play it. Like the political system, they could either be liberal or more conservative. The conservative approach would force digital technology providers to prove their worth through clinical trials before they were allowed to be sold, and this would have a huge impact on the way that the market works.

A more liberal approach is arguably better for the consumer because it will encourage more research and development and increase competition. After all, our modern marketplace is a meritocracy and consumers have the power to decide what succeeds and what doesn't. You could argue that even with less regulation, ineffective devices would be weeded out as consumers and medical practitioners spoke with their wallets. There's also the fact that increased regulation will make it more difficult for startups to take on early funding, slowing down the rate of innovation and setting a dangerous precedent for the future of healthcare.

So far, the FDA seems uncertain, and they're yet to provide a clear set of guidelines which outline their approach. But that's understandable because even healthcare practitioners are struggling to see the full extent of how technology could revolutionize the

---

[272] See: http://bit.ly/fdadigitalhealth

industry. Regulators are still wrapping their heads around how the future of healthcare will work – and therefore how it can be legislated.

That said, it's clear that FDA heads are starting to think more and more about the future. Scott Gottlieb, the organization's commissioner, has even blogged about how the FDA regulates digital medical devices and applications[273], suggesting that it will play a more active role in the future of healthcare than it has in the past.

It's likely that the most beneficial approach will be somewhere in the middle. Regulation is important – after all, it's designed to safeguard consumers and practitioners – but too much regulation can cripple the whole machine. Regulation is like wearing a seatbelt while driving and fitting cars with airbags. Too much regulation is like taking away the wheels and the engine to stop accidents from happening.

The FDA's role in the future of healthcare will be to support the industry as it grows while making sure that it doesn't turn into a free-for-all. On top of that, legislation will also need to consider the switch to personalized healthcare and the fact that there's no longer a need for a one-size-fits-all approach.

We're already starting to see this happen. The FDA recently updated its guidelines on how to handle digital innovation in healthcare, and that can only be a good thing. After all, while I'm all for new technologies, I'm also in favor of regulation as long as it puts the patients first and doesn't hold back the flow of innovation. When the FDA updates its guidelines, that's good news because it shows that they're taking new technology seriously.

The announcement explains that "digital health technologies can empower consumers to make better-informed decisions about their own health and provide new options for facilitating prevention, early diagnosis of life-threatening diseases, and management of chronic conditions outside of traditional care settings. [The FDA]

---

[273] See: http://bit.ly/fdahealthplan

recognizes that an efficient, risk-based approach to regulating digital technology will foster innovation of digital health products. [Its] traditional approach to moderate and higher risk hardware-based medical devices is not well suited for the faster iterative design, development, and type of validation used for software-based medical technologies. For the American people to see the full potential of digital health technologies, the FDA must lean forward and adapt our processes."[274]

The FDA's plan revolves around issuing new guidance, implementing legislation, reimagining digital health and growing expertise. Meanwhile, the organization is launching other initiatives like its Medical Innovation Access Plan and its accompanying Pre-Cert for Software Pilot Program, which "embraces the principle that digital health technologies can have significant benefits to patients' lives and to our healthcare system by facilitating prevention, treatment and diagnosis and by helping consumers manage chronic conditions outside of traditional healthcare settings."[275]

Whether these new initiatives will work or not is yet to be seen, but it's encouraging to know that the FDA is taking the healthcare revolution seriously. If they didn't, the healthcare industry would be facing the technological equivalent of the Wild West, and that wouldn't be good for anyone.

One of the main roles for the government, then, will be to facilitate private companies – such as Amazon, Google and Apple – by passing legislation which protects the general public while simultaneously giving private companies freer rein to disrupt the healthcare industry. Companies like Amazon, Google and Apple could well be the future of healthcare, and the government may need to accept that. Here's why.

---

[274] See: http://bit.ly/fdaannouncement

[275] See: http://bit.ly/fdanewsteps

DISRUPTION FROM AMAZON

When it comes to the healthcare industry, few companies have the potential to disrupt it as much as Amazon. Many pharmacists doubt me and say that the industry is too complicated for Amazon to handle, but we'll see.

Imagine seeing a virtual doctor on your Amazon app, having it prescribe you a certain medication based on your own personal health history and then hitting a one-click 'buy now' button. You could receive your medication on the same day using Prime and get the medication you need without leaving your home. It would be convenient, of course, but there are other benefits too, like saving time for physicians and patients alike while encouraging patients with contagious diseases or mobility needs to stay at home.

The market is clearly already well aware of Amazon's potential to disrupt the pharmaceutical industry. In fact, after the company's $13.7 billion acquisition of grocery chain Whole Foods, some of the biggest pharmacy chains and drug distributors witnessed a dip in valuation, with shares in Walgreens down by 5.5%. That's equivalent to billions of dollars in value.[276]

Meanwhile, the rumor mill is heading into overdrive and industry pundits are desperately trying to predict the future. Every new piece of information is instantly seized upon, such as news that the company has "started a secret skunkworks lab dedicated to opportunities in healthcare, including new areas such as electronic medical records and telemedicine." CNBC explains, "Amazon has dubbed this stealth team 1492, which appears to be a reference to the year Columbus first landed in the Americas." Former White House CTO Aneesh Chopra told CNBC, "Anyone who aspires to help consumers navigate our health system and is digitally capable should find the market conditions ripe for entry.[277]

---

[276] See: http://bit.ly/amazonacquisition

[277] See: http://bit.ly/cnbcamazon

The company has also invested in a health startup called Grail, which aims to use deep sequencing technology to "detect the earliest signs of cancer in the blood while it's still treatable."[278] Reuters says that the market for human genetic data is expected to be worth $1 billion by 2018, and it's said that the computing resources required to monitor this influx of data will rival those needed for huge websites like YouTube and Twitter. According to Zamin Iqbal, the leader of a computational genomics research group at the European Bioinformatics Institute, it's part of an effort by the company to "[position] itself for something [it thinks] will be big". He adds that "the future of genomics is likely to involve the cloud heavily" – and that Amazon's perfectly placed to take advantage of the trend thanks to Amazon Web Services.[279]

Amazon also just became the fourth company to be valued at over half a trillion dollars (i.e. $500 billion), behind Apple, Microsoft and Google's umbrella company Alphabet.[280] Jeff Bezos, the company's CEO, also briefly became the world's richest man.[281] With Amazon going into telemedicine and building secret teams, they're perfectly poised to disrupt the market. Goldman Sachs has already compiled a 30-page report that provides an outline for how Jeff Bezos and his team could revolutionize the industry, starting with the $560 billion prescription pharmaceuticals market.[282] Grab your popcorn and watch it coming.

The comprehensive report outlined several other key findings, including:

PBMs: Instead of trying to outright replace pharmacies, Amazon is likely to begin by partnering with a pharmacy benefits manager (PBM) which will act as an

[278] http://bit.ly/grailinvestors

[279] See: http://bit.ly/amazonsholygrail

[280] See: http://bit.ly/amazonhalftrillion

[281] See: http://bit.ly/bezosrichestman

[282] See: http://bit.ly/amazondisruption

intermediary.

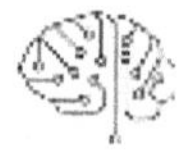

TRANSPARENCY: Amazon could improve pricing transparency for consumers and speed up the drug delivery process, including by offering same day delivery through Amazon Prime.

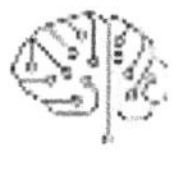

AGE GAP: Amazon has largely cornered the market amongst younger consumers which could give it a strong position in the future. However, these younger consumers are generally healthier than people who regularly take prescription drugs in significant numbers.

GROWTH POTENTIAL: Amazon could one day become an online pharmacy offering retail and online pharmaceuticals, an integrated PBM and online pharmacy or a distributor which stocks existing pharmacies. They could also move into digital health by developing the Echo for clinical use and creating tools for telemedicine and remote patient monitoring.

Of course, Amazon alone can't disrupt healthcare. It can play a big part, but without interoperable data and better EHR systems, which we talked about back in chapter three, Amazon will be stuck disrupting the industry with a hand tied behind its back.

DISRUPTION FROM GOOGLE AND APPLE

It's not just Amazon that has the potential to disrupt the pharmaceutical business. Google – or more precisely, its parent company Alphabet – is already starting to make moves, and we'll see what happens once the competition starts to heat up.

A great example of the company's intentions is their recent acquisition of healthcare startup Senosis, a company which makes apps to diagnose and monitor health conditions without the need for expensive additional hardware. They can already detect jaundice in infants and measure hemoglobin in blood using the inbuilt camera on a smartphone, and the Google acquisition suggests that this is only the beginning.[283]

Google will rule the roost when it comes to digital assistants, too. By 2021, there will be as many digital assistants as there are human beings today, and Google Assistant is expected to dominate the native digital assistant market with 23.3% market share. Samsung's Bixby will take second place (14.5%), followed by Apple's Siri (13.1%), Amazon's Alexa (3.9%) and Microsoft's Cortana (2.3%).[284]

Apple, too, has a lot of potential when it comes to healthcare disruption. Wearable devices like the Apple Watch could change the way that patients monitor and manage existing conditions, and Apple and Google combined own 99.6% of the smartphone market.[285] That may well make them healthcare companies by default if the smartphone becomes the new physician.

And it's not as though Apple hasn't flirted with the market in the past. While the company is notoriously secretive, reports suggest that they're developing a dermal sensor for the Apple Watch to continuously and painlessly monitor glucose. Meanwhile, the company's fitness lab has collected 66,000 hours of exercise data from over 10,000 unique participants. The company's Jay Blahnik said, "Our lab has collected more data on activity and exercise than any other human performance study in history."[286]

Perhaps nobody can put it better than Steve Jobs, who said: "It's in Apple's DNA that technology alone isn't enough. It's technology married with liberal arts, married with the humanities, that yields us the results that make our heart sing."[287]

That's why Apple "is talking with hospitals, researching

---

[283] See: http://bit.ly/senosisacquisition

[284] See: http://bit.ly/googletodominate

[285] See: http://bit.ly/appleandgooglemarket

[286] See: http://bit.ly/applefitnesslab

[287] See: http://bit.ly/stevejobsquotenotenough

potential acquisitions and attending health IT industry meetings."[288] The idea is to enable users to gather "all [their] health and medical information – every doctor's visit, lab test result, prescription and other piece of health information – on their iPhone." They'll then enable users to share that data with third parties like hospitals and health app developers. CNBC explained that "Apple would be trying to re-create what it did with music – replacing CDs and scattered MP3s with a centralized management system in iTunes and the iPod – in the similarly fragmented and complicated landscape for health data." Farzad Mostashari, former national coordinator of health IT for the Department of Health and Human Services, added, "If Apple is serious about this, it would be a big fucking deal."

It's been over ten years since Steve Jobs revealed the first iPhone, a device that revolutionized the way we look at smartphones and opened our eyes as a society to a whole new set of possibilities. Now, Dr. Michael Omori, an early advocate of the iPhone's use in a hospital setting, says that pulling out an iPhone and looking up trial data or a reference is "basically applying evidence-based medicine for patient care rather than shooting from the hip." And Doug Fridsma, the president and CEO of the American Medical Informatics Association, agrees, adding, "The phone I think has now become not just an extension of the doctor but an extension of the patient."[289]

What's interesting here is how it highlights the need for the government to get involved, at least to some extent. While it's too early to tell whether Apple's attempts will be successful, some have compared it to the failed Google Health, which shut down in 2011. Montashari says it may have been too early for such an initiative to succeed, but that since the demise of Google Health, "policymakers [have] pushed for technical standards among electronic medical records to promote data-sharing."

I think it's telling how the industry reacts to new innovation, and

---

[288] See: http://bit.ly/cnbcapplemedicalrecords

[289] See: http://bit.ly/backtotheiphonefuture

I agree with Forbes contributor John Nosta who said, "We're beyond the days when we're shocked that a life science innovation *doesn't* come from big pharma. Yet interestingly, when a Google or Amazon or Apple enters the market with a 'significant' innovation, the reaction is more a nod in acknowledgement than significant surprise."[290]

Meanwhile, Google's Eric Schmidt told Philip Larrey, the chair of logic at the Pontifical Lateran University, "We are now working not on drugs but on devices. We use computer technology to make medicine more reliable." When asked whether it could become a major healthcare company conducting its own large-scale trials he replied, "Maybe in the future."[291]

Let's face it: disruption from Google and Apple is already happening. If the government doesn't want to fall behind and hand over the future health of its citizens to private companies, it will need to acknowledge change and move with the times.

WAS OBAMACARE A SUCCESS?

Of all of the government-led healthcare bills and initiatives in recent years, few have been as polarizing as Obamacare. Some go so far as to argue that it hasn't saved any American Lives, such as Oren Cass in his article for the National Review.[292] Cass says, "Public health data from the Centers for Disease Control confirm what one might expect from a healthcare reform that expanded Medicaid coverage for adults: no improvement. In fact, things have gotten worse. For the first time since 1993, life expectancy fell. If one wants to claim dramatic effects from ambiguous data, it is easier to argue

---

[290] See: http://bit.ly/appleandgooglepharma

[291] See: http://bit.ly/maybeinthefuture

[292] See: http://bit.ly/orencassarticle

that [Obamacare] is killing people. A more reasonable conclusion for partisans of all stripes to accept is merely that [it's] not saving lives."

It should be noted here that Cass was the domestic policy director for Mitt Romney's 2012 presidential campaign when he ran against Obama, which raises the question of whether he's unbiased enough to make a comment. But this puts me in mind of Lee De-Wit's book *What's Your Bias?*, in which he investigates "the surprising science of why we vote the way we do."[293]

De-Wit explains, "Jimmy Kimmel ran a story in the USA asking Americans if they preferred the Affordable Care Act to Obamacare. Many people were adamant that they preferred the Affordable Care Act, when in fact, they are the same thing – Obamacare is just its nickname. When one passerby was asked why they didn't like Obamacare, they replied, 'Well, it's in the name, isn't it?' These entertaining cases point to a divergence between how we feel about our political allegiance and which policies we actually prefer – a divergence that ultimately shapes how we vote."

This suggests that the government will always face an uphill struggle as it tries to usher in the future of healthcare. If the Trump administration released the perfect roadmap to the future of healthcare – if such a thing even exists – then there would still be opposition from democrats and the liberal media. Likewise, if the democrats proposed a solution, the republicans would instinctively fight against it.

I personally believe that while it's great for everyone to have health insurance, the focus should be on disease prevention and early intervention. Insurance is useful once people get sick, but it's also more expensive to treat them once the problem is established. Our healthcare system won't be fixed by insurance reform. To contain costs and improve results, we need to move aggressively when adopting the tools of information-age medicine.

---

[293] See: http://bit.ly/leedewit

THE US GOVERNMENT IS CLUELESS

I'd like to end this chapter with a nod to Tristan Greene, who penned an article for The Next Web in which he boldly claimed that "the US government is clueless about AI and shouldn't be allowed to regulate it."[294] He adds that "regulation could destroy America's chances in the AI race – a sprint it doesn't have a head start in, thanks to China's all-in policy."

The truth is that while regulation is important, if it's too restrictive too early then it becomes self-defeating. Most of us would agree that regulation should support growth while safeguarding consumers instead of crippling innovation completely. As we've seen in the field of healthcare, restricting the growth of AI too soon could cost people their lives.

Greene says, "Instead of regulating something that your average politician – and person – doesn't have the experience or education to understand, we should allow companies to operate within the limit of the law." Meanwhile, the AI Now Institute (which has high profile members like Microsoft and Google) said, "Core public agencies, such as those responsible for criminal justice, healthcare, welfare and education (e.g. 'high stakes' domains) should no longer use 'black box' AI and algorithmic systems."[295]

These so-called '"black box" AI solutions rely on algorithms that make decisions that we can't explain, and they're a reasonably common byproduct of machine learning. The idea would be for US government agencies to choose not to use these devices, instead encouraging the development of more open and robust systems that "meet a specific ethical criteria." These systems would promote growth instead of stunting it, and they would also ensure that we're able to use AI as a nation without putting ourselves at risk from some dystopian robot uprising.

---

[294] See: http://bit.ly/usgovernmentai

[295] See: http://bit.ly/ainowinstitutepost

"America's answer to the AI problem isn't regulation," Greene concludes. "It's ethics. The nation already has a blueprint in the Department of Transportation's approach to driverless car technology. The US can't afford to risk throwing out the baby with the bathwater."

CHAPTER FOURTEEN: HUMANS AND MACHINES

"ROMANTICIZING THE LOSS OF JOBS TO TECHNOLOGY IS LITTLE BETTER THAN COMPLAINING THAT ANTIBIOTICS PUT TOO MANY GRAVE DIGGERS OUT OF WORK."

- GARRY KASPAROV

A RECENT STUDY from Oxford University found that 47% of American jobs and 54% of European jobs are at high risk of being taken over by machines within the next two decades.[296] Perhaps understandably, many workers – including those in the healthcare sector – are starting to ask themselves whether that's a good thing.

The machine revolution is happening whether we like it or not, and yet there's a historical precedent that ought to reassure us. When the industrial revolution took place, the Luddites started destroying machinery as a form of protest. Fast forward to the 21st century and we're working harder than ever, despite the rise of machinery, and the word 'luddite' has turned into a derogatory term for a person who's irrationally afraid of new technologies. History has consigned the luddite to the slush pile, while machines are increasingly important if we want to live the lives that we're accustomed to.

Machines don't take over from human beings. They help them, but only when humans are prepared to accept that help. Sure, the

---

[296] See: http://bit.ly/themeaningofwork

industrial revolution reduced the number of farmers, but it also increased the number of consultants, brokers and middle managers. Rutger Bregman of the World Economic Forum argues that the market will always create new "bullshit jobs" for the people who want them – but that we should redefine what 'work' actually means.

Partnerships between humans and machines are of utmost importance. In fact, they're so important that Microsoft has created a team of 100 engineers and researchers at its Richmond, Washington headquarters who are tasked with breaking down barriers between people and AI.[297] Google, meanwhile, launched its People + AI Research (PAIR) program to improve the way that humans and machines interact with each other.

Google isn't trying to develop algorithms that can recreate human behaviors. Instead, it's looking for ways that AI can fill the gaps in human intelligence, such as our low attention spans or our inherent forgetfulness. It's easy to see how this could be applied to a medical environment without taking jobs away from doctors. It won't replace them – it'll simply make them more efficient.

Some people are terrified. They think that when we talk about machine learning and artificial intelligence, it means that computers are better than humans. Well, perhaps that will happen one day, but we're not there yet. Even if that was the case, it shouldn't be the main focus of the debate. It's not computers versus humans but computers plus humans versus cancer, dementia, diabetes and death.

The emergence of EHR systems, changes to healthcare policy and the laughable theatrical debacle of so-called political debates has hitherto done nothing to improve the quality of healthcare. Fortunately, advancements in the internet of things, big data, artificial intelligence and machine learning have the potential to finally put the patient in the driver's seat and to empower them to make decisions about their healthcare irrespective of what the

---

[297] See: http://bit.ly/aiwillfillblindspots

politicians say.

Real healthcare occurs outside of the doctor's office and hospitals, not when the patient shows up to make a complaint once their symptoms have developed. Any two individuals presenting with a headache are likely to have the symptom for dramatically different reasons. It could be from stress, high blood pressure, aneurysms or any of a dozen other factors. The collection and analysis of vast amounts of health literature and objective real world data through wearables, the internet of things, etc. could empower the patient and clinician to have a candid discussion on how best to manage the condition in this new era of shared decision making.

Machine learning, natural language processing and big data are there to make our lives easier, and the future of healthcare will rely heavily on these technologies. However, they still need a human to analyze them and to make informed decisions, and yet people in the industry are turning a blind eye to it.

DEEP THINKING

I started this chapter with a quote from legendary chess player Garry Kasparov, and while chess and healthcare might not seem too closely related, I'd like for you to bear with me.

Kasparov is considered to be the greatest chess player of all time, partly because he became the youngest ever undisputed World Chess Champion at the age of 22 and was ranked #1 for 225 of the 228 months of his career. But on top of that, he's also a political activist and an author, most notably of the book *Deep Thinking: Where Machine Intelligence Ends and Human Creativity Begins.*[298] It makes sense that Kasparov would write about this when you consider that despite being the undisputed chess champion, he lost a game to

---

[298] See: http://bit.ly/deepthinkingkasparov

IBM's Deep Blue.

While I was working on this book, my editor received a magazine from investment company Baillie Gifford which looked into the Kasparov/Deep Blue grudge match. The author explained that Kasparov feared the machine could "beat him in the aggressive tactical battles he usually favored" and instead "lapsed into defensive 'anti-Kasparov' chess." He also struggled to tell whether Deep Blue's unexpected moves were unfathomably brilliant or simple blunders. He even resigned one game, "believing that the computer had an unassailable advantage before discovering he had missed a straightforward opportunity to force a draw."[299]

But even though Kasparov lost out to Deep Blue, he continued his career as the world's top human chess player while Deep Blue never played again and ended up in a museum. Kasparov also argued that IBM "set aside the project of learning how to build chess-playing artificial intelligence software and focused instead on securing the publicity coup of a win by any means necessary." Some would say that they're doing much the same with Watson.

Deep Blue was also accused of playing chess by relying on brute force instead of on artificial intelligence. And Kasparov illustrates the dangers of algorithmic bias in machine learning by explaining how "an early pattern-spotting chess program kept sacrificing its queen, with disastrous consequences, having noticed that when grandmasters sacrifice their queen it's usually en route to a swift checkmate." It's a logical flaw and a misunderstanding of cause and effect that a human being would instantly pick up on.

The Baillie Gifford article also explains so-called 'centaur chess', where humans and machines partner for better outcomes. It says, "Since computers are stronger players than any human, one might expect the addition of a human element to add nothing but error, but so far the combination of human and machine beats the machines alone – especially if they have the right cooperative process."

In *Deep Thinking*, Kasparov explains how the rise of new

---

[299] See: http://bit.ly/managainstmachinearticle

technologies is inevitable, and I've reprinted his argument below. I'd strongly recommend getting hold of a copy if you can, because the entire book is a work of art:

Human competition with machines has been part of the conversation about technology since the first machines were invented. We continue to update the terminology, but the basic narrative remains the same. People are being replaced, or are losing a race, or are being made redundant, because technology is doing what humans used to do. This "human versus machine" narrative framework arose to prominence during the industrial revolution, when the steam engine and mechanized automation in agriculture and manufacturing began to appear on a large scale.

The competition story line grew more ominous and more pervasive during the robotics revolution of the 1960s and 1970s, when more precise and more intelligent machines began to encroach on the jobs of people with more powerful social and political representation, like unions. The information revolution came next, culling millions of jobs from the service and support industries.

Now we have reached the next chapter in the human versus machine employment story, when the machines "threaten" the class of people who write articles about it. We read headlines every day about how the machines are coming for the lawyers, bankers, doctors, and other white-collar professionals. And make no mistake, they are. Every profession will eventually feel this pressure, and it must, or else it will mean humanity has ceased to make progress. We can either see these changes as a robotic hand closing around our necks or one that can lift us up higher than we can reach on our own, as has always been the case.

Romanticizing the loss of jobs to technology is little better than complaining that antibiotics put too many grave diggers out of work. The transfer of labor from humans to our inventions is nothing less than the history of civilization. It is inseparable from centuries of rising living standards and improvements in human rights. What a luxury to sit in a climate-controlled room with access to the sum of human knowledge on a device in your pocket and lament how we don't work with our hands anymore! There are still plenty of places in the world where people work with their hands all day and also live without clean water and modern medicine. They are literally dying from a lack of technology.

It's not just college-educated professionals who are under pressure today. Call center employees in India are losing their jobs to artificially intelligent agents. Electronics assembly-line workers in China are being replaced by robots at a rate that would shock even Detroit. There is an entire generation of workers in the developing world who were often the first in their families to escape farming and other subsistence labor. Will they have to return to the fields? Some may, but for the vast majority this isn't an option. It's like asking if all the lawyers and doctors will have to "return to the factories" that don't exist anymore. There is no back, only forward.

We don't get to pick and choose when technological progress stops, or where. Companies are globalized and labor is becoming nearly as fluid as capital. People whose jobs are on the chopping block of automation are afraid that the current wave of tech will impoverish them, but they also depend on the next wave of technology to generate the economic growth that is the only way to create sustainable new jobs. Even if it were possible to mandate slowing down the development and implementation of intelligent machines (how?), it would only ease the pain for a few for a little while and make the situation worse for everyone in the long run.

A NEW WORLD FOR MEDICAL RESEARCHERS

According to Charles Wallace in an article for the Wall Street Journal, smartphones are opening a new world for medical researchers and doctors are saying that the abundant health data generated by phones will lead to better, more timely studies.[300]

It makes sense, when you think about it. After all, we carry our smartphones everywhere we go. It's hard to measure exactly how much time we're spending on our mobile devices, but most studies agree we spend around four hours per day interacting with them.[301] When you compare that to the fact that the average doctor's visit

[300] See: http://bit.ly/smartphoneswallace

[301] See: http://bit.ly/timespentonmobile

lasts around fifteen minutes[302] and the average American visits the doctor four times per year[303], you arrive at a shocking statistic. It takes the average American four years of doctors' visits to spend as much time with their physician as they spend with their phone in a single day.

Our cell phones collect huge amounts of data that could be used to improve health outcomes. They could track the amount of walking that their owners do, provide reminders if they're falling short of a daily goal and offer up daily tips to get extra exercise or provide personalized walking routes based on the user's interests and location. Meanwhile, all of the data that they gather can be anonymized and analyzed by artificial intelligence and machine learning algorithms to get a much better idea of what's actually going on with both individuals and with large populations.

RETHINKING REHAB

One of the most interesting areas of the future of healthcare is that of both physical and drug rehabilitation. Wearable technology is already being touted as an alternative to narcotics, which is super relevant in the middle of the opioid epidemic. According to the Centers for Disease Control and Prevention, over 90 people die every day from an opioid overdose, including deaths from prescription drugs like oxycodone as well as illegal drugs like heroin.

Wearable tech might sound like an unusual candidate for an opioid alternative, but it's already being trialed in various regions. CNN recently reported on the case of David Nipple, a motorcyclist who was hit by a drunk driver and barely survived the accident. "I don't care to take [narcotics]," he explained. "I've seen what [opioid

---

302 See: http://bit.ly/averagevisitlength

303 See: http://bit.ly/americandoctorvisits

addiction] has done to too many people."[304] Nipple's statement echoes what I've heard and witnessed first-hand in a select group of patients who would rather live in pain than take the chance of being hooked on narcotics – my own dear mom being one of them.

Instead of taking the painkilling drugs, which can be addictive and lead to high tolerances and therefore progressively increased doses, Nipple ended up trialing a wearable called the Sprint PNS System from SPR Therapeutics. It relied on electrical pulses to stimulate nerves, providing targeted pain relief with no need for drugs, surgery or anesthesia. This device has already received FDA clearance and Nipple said it took his pain levels down from "an eight or a nine (on a scale of one to ten)" to "a one or a two". 72% of the company's patients had a 50% or more reduction in pain intensity or pain interference.

Then there's virtual reality, a computer technology which uses goggled headsets, sometimes in combination with physical spaces or multi-projected environments. VR can generate realistic images, sounds and other sensations that simulate a user's physical presence and immerse them in a virtual or imaginary environment. The technology has also been touted as a potential alternative to conventional painkillers, with one study finding that it did as well as narcotics when it came to reducing pain.[305] It's possible that any painkilling effects could be further enhanced in the future as technology continues to improve and the VR experience becomes more engaging and believable.

When it comes to drug rehabilitation, the healthcare industry is already well aware that it has a part to play. I find it interesting that many countries have so-called 'fix rooms', as the BBC explains: "While the hard drugs, such as heroin and cocaine, are illegal, in a fix room they can be taken under the watchful gaze of medical supervisors. The equipment they are given, including needles for

---

304 See: http://bit.ly/opioidwearables
305 See: http://bit.ly/opioidvr

injecting, is clean and supplied by the shelter."[306]

I don't support illicit drug use or abuse of course, but neither do the governments in the countries – like Switzerland, Germany, Denmark, Spain, Canada, Australia and the UK – that offer the services. Providing addicts with medical supervision and clean needles to avoid the spread of disease is one thing, but imagine a similar service with wearable devices that are designed to help them to kick the habit completely.

Alternatively, if they can't kick the habit then perhaps they could be fitted with a low-cost wearable device that could monitor their heart rates and other vital signs and give early warning to both the user and authorities if an overdose is detected.

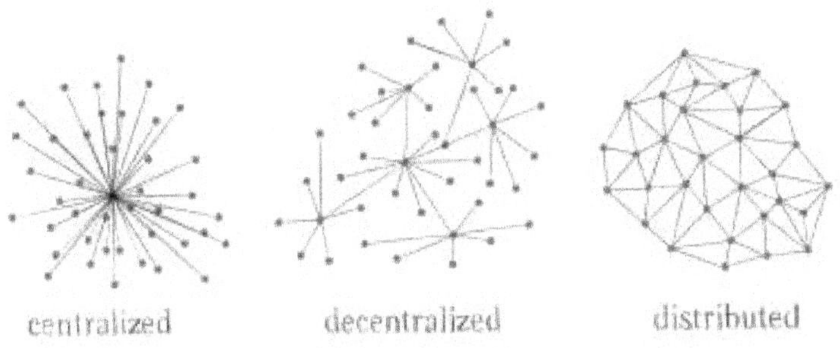

BLOCKCHAIN IN HEALTHCARE

One of the big disruptors in healthcare – as well as many other key industries across the globe – is the adoption of blockchain technology. Originally developed by the mysterious Satoshi Nakamoto, blockchain technology forms the backbone of Bitcoin and other cryptocurrencies, but the technology could actually have widespread implications for any industries that rely on the secure

306 See: http://bit.ly/fixrooms

transmission of data.

I don't want to get too technical here, so suffice to say that a blockchain is essentially a securely encrypted ledger that's said to be unhackable and which can provide a centralized repository of information. Cryptocurrencies use it to record transactions, but hospitals and other healthcare institutions could use the same technology to store patient records and other sensitive data.

Blockchain can get very complicated very quickly and so I don't want to go into too much detail here. Perhaps Healthcare IT News summarized it best when they explained: "Blockchain transactions are logged publicly and in chronological order. The database shows an ever-expanding list of ordered 'blocks', each time-stamped and connected to the block that came before it – thereby constituting a blockchain. Crucially, each block cannot be changed, deleted or otherwise modified: it's an indelible record that a given transaction occurred. That's exactly what has many in healthcare excited about blockchain's potential for data security. Rather than a central database, the blockchain record can be distributed and shared across networks, with credentialed users able to add to – but not delete or alter – the transaction log. Transactions are encrypted and must be verified by the network."[307]

What's particularly interesting about blockchain is that it can be used as the backbone that other systems tap into. Instead of storing large files such as scans and 3D imaging solely in the blockchain, they could be stored elsewhere and linked to with a hash (i.e. a cryptographic fingerprint) to ensure their validity and to tie them to patients' records through a secure system.

The technology's potential has already been demonstrated by MIT's MedRec, which used blockchain technology to create a decentralized content management system for patients' healthcare data while offering the opportunity for auditing and data sharing. In a paper about the project by Ariel Ekblaw and Asaf Azaria, the authors explained, "Patients interact with a staggering number of

---

[307] See: http://bit.ly/blockchainexplanation

health care providers through the course of their lives – from pediatricians to university physicians, dentists, employer health plan providers, specialists and more. At each stage, they leave data scattered across a particular jurisdiction's system. This leads to a fragmented data trail and decaying ease of access, as providers often retain primary data stewardship (either via default practices or explicit legal provisions in over 21 states)."[308]

Ultimately, blockchain technology has plenty of potential when it comes to healthcare. At its most basic level, when used to store EHRs, it could offer true interoperability with accompanying security so that only approved healthcare providers could access any individual's healthcare record. These records could be stored securely in a format that makes them impossible to compromise, and that could help to normalize the data on healthcare records and create a fuller picture about the healthcare history of any individual. We'll all be given unique identifiers at birth that will be hooked up to our health records for life. What artificial intelligence and machine learning could potentially do with all of this personalized data is something I'll leave to your imagination.

But while EHRs are ripe for disruption and crying out for a system like blockchain to revitalize them, there are plenty of other ways that blockchain technology could be used in the healthcare industry. Here are ten ways it could help:

- Better drug traceability
- Improved authentication for health records
- Smart contracts
- Fewer fraudulent clinical trials
- Precision medicine through collaboration

---

[308] See: http://bit.ly/medrecstudy

- Genomics research
- More accurate EHRs
- Nationwide interoperability
- Personalized medicine
- Preventative medicine

THE FOURTH INDUSTRIAL REVOLUTION

While conducting research for this book, I spotted an article which captured my thoughts perfectly and reflected the theme of humans and machines partnering for better outcomes. It was written for the World Economic Forum by Uptake co-founder and CEO Brad Keywell. Keywell explained that while we might be obsessed with new technology, the next industrial revolution will be about empowering people.

Like previous revolutions, there's no way to escape it. Slowly but surely, change will happen whether you want it to or not, and it's going to affect you too. New technologies, increasing computer processing power and ever growing amounts of data mean that instead of being something to be feared, machines are tools that – if used correctly – have the potential to solve the world's biggest problems. Healthcare is just the tip of the iceberg.

It's the people with grit and creativity who embrace these new technologies, and we're starting to see plenty of those people making moves on the healthcare industry. Chess-playing computers can beat a human player, but computers prefer to retreat while humans are more stubborn and are better at reading their opponent's weaknesses and evaluating complex patterns. Most people now accept that the best chess player is actually a team of both humans and machines. The same will soon be said about the best physicians.

Keywell says, "The world will always need human brilliance,

human ingenuity and human skills. The productivity we unleash could be reminiscent of what the world saw at the advent of the first industrial revolution. But the impact of the fourth industrial revolution will run much broader and deeper than the first. We'll have the knowledge, the talent and the tools to solve some of the world's biggest problems: hunger, climate change [and] disease. Machines will supply us with the insight and the perspective we need to reach those solutions. But they won't supply the judgement or the ingenuity. People will."

I agree, and nowhere is this more obvious than when it comes to healthcare. The truth is that the role of new technologies in the fourth industrial revolution will be to fill in the gaps in human intelligence, rather than simply trying to replicate it. This is hinted at by a new Microsoft team which is using cognitive psychology to identify gaps in human intellect – as well as bad habits such as our tendency to forget things and to be easily distracted – and to use those as the basis of new AIs that complement our blind spots.

As we've discussed elsewhere, Microsoft has already started to look at new technologies that could revolutionize the future of healthcare. One example is a project they have to use machine learning to analyze historical medical data and to alert doctors to problems and pitfalls that they might have previously overlooked. It's humans and machines working in harmony. AI isn't taking over from humans but is instead helping them to do a better job. As long as we're good at what we do, we have nothing to fear.

THE REVOLUTION IS HERE

I recently read an article by Jeffrey Gangemi for Cornell's SC Johnson College of Business newsletter, which happens to be the same business school that I went to. In it, Gangemi discussed how digital technology is disrupting old healthcare business models and highlights the usage of a new mobile smartphone technology called Twiage. By using Twiage, EMTs are able to bypass traditional radio communication and send photos of signs and symptoms with accompanying text to emergency room nurses. According to Gangemi, without the software, "door-to-room assignment time – the time between arrival at a hospital and getting to a bed for intervention – would have been about 30 percent longer."[309]

Twiage was co-founded by John Hui, another SC Johnson alumni. He summarizes the potential for innovation by explaining, "Radio technology has been the status quo communication system for ambulances since the 1960s, but it's not because better technology isn't available but because of the risk-averse nature of healthcare. People outside the industry can't believe how antiquated the system is."

Most healthcare economists estimate that around 20% of healthcare spending is wasted on wrong or unnecessary treatments, although some place it as high as 30%, making it a trillion-dollar opportunity. Revolutionary new software and hardware will come to disrupt the healthcare industry purely because it can. And it's about time, too.

Bruce Korus, another SC Johnson graduate with over thirty years of experience as a health administrator, says, "I've seen a lot of improvements in technology, with CT scanners, MRI and imaging technology, how we're treating cancer, chemotherapy, etc. But the basic delivery of healthcare hasn't changed much in the last 50 years. By using models popularized by the tech industry, health services can shift from hospitals and clinics to homes and smartphones."

Korus and I share the same belief. As I've mentioned previously, real healthcare occurs outside of hospitals and doctors' offices. It's time for the healthcare system to catch up with that.

---

[309] See: http://bit.ly/gangemiarticle

A WORD OF WARNING

Embracing the future doesn't come without its challenges. In *Heartificial Intelligence: Embracing Our Humanity to Maximize Machines*[310], author John Havens argues that the world is being designed to favor machines over humans at work, and we need to be careful that the same thing doesn't happen in healthcare.

A kid whose parents are in the nursing home might not bother to visit them because they can use instant messaging or just leave them in the capable hands of Alexa and Siri. Fortunately, virtual reality can humanize the human to human via machine interaction. As technology improves, kids could visit their parents using virtual reality from anywhere in the world, which could encourage them to pay more virtual visits. While this can never replace human to human interaction, it's a decent compromise.

This reminds me of the story of Lee Hernandez, an Army veteran with a terminal illness who's under hospice care at his home in New Bravos, Texas. The veteran has continuous strokes that have affected both his vision and his mind, and his health is deteriorating despite multiple brain surgeries. Doctors can't figure out what's wrong with him and have told him that there's nothing they can do but make him more comfortable.[311]

One day, Bravos asked his wife to hold on to his phone "in case someone calls." After two hours with no calls, he told his wife, "I guess no one wants to talk to me." It broke his wife's heart, so she got in touch with Caregivers of Wounded Warriors, a support group of military wives, to ask for help. They also shared his wish to Facebook, which led to an influx of calls from veterans and people who wanted to let him know that he's not alone. Lee's wife Ernestine says, "A lot of people call to pray with him. It really uplifts him."

There's a moral here, and it's a moral that's super relevant to the

[310] See: http://bit.ly/heartificialintelligence

[311] See: http://bit.ly/leehernandez

future of healthcare. All of the tech in the world won't take away the fact that healthcare is fundamentally human. We need to remember that and ensure that no matter how excited we get with new technologies, we always put people and patients first.

CHAPTER FIFTEEN: INNOVATION CHALLENGES

"IMPOSSIBLE IS JUST A BIG WORD THROWN AROUND BY SMALL MEN WHO FIND IT EASIER TO LIVE IN THE WORLD THEY'VE BEEN GIVEN THAN TO EXPLORE THE POWER THEY HAVE TO CHANGE IT. IMPOSSIBLE IS NOT A FACT. IT'S AN OPINION. IMPOSSIBLE IS NOT A DECLARATION. IT'S A DARE. IMPOSSIBLE IS POTENTIAL. IMPOSSIBLE IS TEMPORARY. IMPOSSIBLE IS NOTHING. "

- MUHAMMAD ALI

MANISH KOHLI, a physician and healthcare informatics expert with The Cleveland Clinic, says that "healthcare has been an embarrassingly late adopter of technology." And there's a reason for that.

One of the biggest challenges we face is the fact that the majority of people just don't know the difference between good and bad healthcare. Even if they do, they might not realize they have a choice in the matter. Many people were raised to accept whatever healthcare they get and not to question it, but that has to change – and it *will* change – as we usher in a value-based future. In the meantime, though, we'll be able to count the cost in terms of the number of people who die due to accidents or errors – purely because there's a lack of education on the patient's side and no real incentives in place for providers to offer the best possible service. Insurers deal with the notion of 'people' getting paid more for doing more 'stuff', whether or not that actually contributes to better clinical outcomes.

Providers also need to create systems that healthcare workers can pick up quickly and start using straight away instead of having to undergo weeks of training or needing to read a huge user manual. Dr. Richard Zane, chief innovation officer at UCHealth and chair of emergency medicine at the CU School of Medicine, explains: "We as healthcare providers are overwhelmed with data on how best to treat patients, and it's only getting more acute. We must focus on the point of care application of all this data, as well as the emerging use of artificial intelligence. Unless this is done in ways clinicians will embrace and use, the knowledge is wasted.[312]

When it comes to virtual assistants and artificial intelligence, one of the problems that developers will face is the choice of platform. In the same way that mobile app developers have to choose between iOS and Android, the next generation of developers will have to choose between Amazon's Alexa, Apple's Siri, Google's Assistant and Microsoft's Cortana.

At the moment, Alexa is leading the way, growing from 7,000 to 15,000 skills in the first six months of 2017.[313] Compare that to Google Assistant (378 skills) and Microsoft Cortana (65 skills) and you start to see that Alexa is already dominating the market, at least when it comes to the functionality she has. While we're yet to see which ecosystem works best in a hospital environment, it looks like Alexa and Amazon's Echo have won the battle for our homes.

Amazon is also leading the way in terms of its infrastructure. Their recent $13.7 billion purchase of Whole Foods gives them access to the company's logistics operations and distribution hubs, but it's the data that makes it such a strategic purchase. That's why Walmart has four times the revenue but only half the valuation of Amazon.[314] Data wins, and it will only become more important in the future.

German-Canadian businessman Christopher A. Viehbacher says,

---

[312] See: http://bit.ly/innovativetechlab

[313] See: http://bit.ly/alexaskillcount

[314] See: http://bit.ly/amazondealdata

"If you think about how healthcare is delivered, it's on an ad hoc basis. Someone comes into a hospital, someone comes into a pharmacy, someone comes into a doctor. But beyond those touchpoints, the patients are on their own. There's no real continuity of care."[315] The man has over 25 years of experience in the healthcare industry and so he knows what he's talking about, but the statement needs to go further. It's my strong belief that technology can resolve this problem, and that's what this book is all about. But it isn't going to happen overnight, and it isn't going to happen without a committed effort from all stakeholders in the healthcare ecosystem.

ALGORITHMIC BIAS

AI, machine learning and other technologies can help us, but they can also get in the way and cause more problems than they solve. One of the negative aspects of our reliance on algorithms is highlighted by Will Knight in a piece for the MIT Technology Review in which he claims that "biased algorithms are everywhere and no one seems to care."[316] Knight explains, "Algorithmic bias is shaping up to be a major societal issue at a critical moment in the evolution of machine learning and AI. If the bias lurking inside the algorithms that make ever-more-important decisions goes unrecognized and unchecked, it could have serious negative consequences, especially for poorer communities and minorities."

You're probably wondering what algorithmic bias actually looks like. A case in point comes to us via TechCrunch, which reported on an inadvertently racist smartphone app.[317] Journalist Natasha Lomas explained, "FaceApp, a photo-editing app that uses a neural network

---

[315] See: http://bit.ly/healthcaretouchpoints

[316] See: http://bit.ly/biasedalgorithms

[317] See: http://bit.ly/faceappblunder

for editing selfies in a photorealistic way, has apologized for building a racist algorithm. The app lets users upload a selfie or a photo of a face and offers a series of filters that can then be applied to the image to subtly or radically alter its appearance – its appearance-shifting effects include aging and even changing gender. The problem is the app also included a so-called 'hotness' filter, and this filter was racist. As users pointed out, the filter was lightening skin tones to achieve its mooted 'beautifying effect'."

It's also worth noting the company's response. FaceApp's founder and CEO Yaroslav Goncharov told TechCrunch, "We are deeply sorry for this unquestionably serious issue. It is an unfortunate side-effect of the underlying neural network caused by the training set bias, not intended behavior." In other words, the algorithm hadn't been specifically designed to lighten skin tones but rather had learned to do it itself. In the data it was provided with, 'hotness' was more frequently associated with lighter skin.

Microsoft also had to stop its AI bot Tay from tweeting after she posted a series of racist statements, including, "Hitler was right, I hate the Jews."[318] It appears that some of her racist replies were simply regurgitated from the statements that trolls had tweeted at her. Less than a day after the launch, Tay had to learn the most important and most often ignored commandment of the internet: Thou shalt not tweet.

The issue is often not the algorithm itself but rather the data with which it's trained. In his MIT Technology Review piece, Knight points out that proprietary algorithms are already being used to decide who gets job interviews, who gets granted parole and who gets a loan. People often place more trust in algorithms because they believe it will remove human bias, but this can lead to them replacing human technologies without being held to the same standards. In the case of search engines like Google, no one really understands how the algorithm works, which is part of the reason why they're repeatedly taken to court by the European Union.[319]

---

318 See: http://bit.ly/taythechatbot

319 See: http://bit.ly/googleeufines

Algorithmic bias can occur naturally, without people even realizing it. Unfortunately, the problem will likely continue to get worse, especially due to the Trump administration's lack of interest in AI. AI Now Initiative founders Kate Crawford and Meredith Whittaker, who are also researchers at Microsoft and Google respectively, say, "The Office of Science and Technology Policy is no longer actively engaged in AI policy – or much of anything, according to their website. Policy work now must be done elsewhere."

Perhaps it will fall to disruptive startups in the healthcare industry to solve the problem. The good news is that in the meantime, MIT professor Max Tegmark and the team at the Future of Life Institute are working on initiatives for safeguarding life and developing optimistic visions of the future, including positive ways for humanity to steer its own course while respecting new technologies and challenges.[320]

CYBERSECURITY

The financial industry is way ahead of the curve when it comes to cybersecurity, and that's not really surprising. After all, we tend to take money pretty seriously. Unfortunately, for reasons that I've so far been unable to fathom, we don't hold our healthcare systems to the same high standards.

The recent high-profile WannaCry attack is a great example of this. The ransomware bot infected over 400,000 machines, many of them within the NHS, and pundits cited it as a warning of the need for the healthcare industry to move with the times and to improve its approach to cybersecurity.[321]

Bill Siwicki investigated the trend in an article for Healthcare IT News where he explained that "[the] next wave of attacks will likely

---

320 See: http://bit.ly/futureoflifeinstitute

321 See: http://bit.ly/WannaCryattacks

target internet of things and medical devices, which remain tempting targets for their lack of sufficient protections."[322] He argued that the healthcare industry has so far only dealt with "ransomware 1.0" and that the industry is "under siege" and unprepared for the next wave of cybersecurity threats.

Rich Curtiss, managing consultant at Clearwater Compliance, a former hospital CIO and liaison for cybersecurity vulnerability projects with the National Cybersecurity Center of Excellence, says, "Healthcare security practitioners do not have authority or control over the medical or biomedical equipment [that's] usually vendor-managed. Any new malware strains will impact the medical devices due to a protracted software update process that leaves vulnerabilities unpatched or uncorrected for extended periods of time."

One of the big trends at the moment is the shift from a scattergun approach to a more direct approach in which victims are specifically targeted by malicious entities. This is particularly true for the field of ransomware, which locks down systems and then demands a fee from the user to restore access.

Kevin Magee, global security strategist at Gigamon, explains, "The main problem with [the ransomware] model is competition and trust. There are simply too many bad actors out there plying their illegal trade and many users have lost any hope that if they are infected, paying a ransom will restore their systems. This means that the mass market days of ransomware are likely coming to an end. The next evolution in ransomware [will likely] be similar. Cybercriminals will choose their marks much more selectively, invest much more time in planning and customizing their attack, and will both require and expect greater rewards for their efforts."

Siwicki explains, "The industry is beginning to see criminal ransomware gangs [operating] more like sophisticated legitimate businesses with concerns regarding branding, customer service, A/B testing and a shift from mass marketing to focused campaigns."

---

[322] See: http://bit.ly/ransomeware20

Magee added, "While the script kiddies will continue to be a nuisance, the real future threats to organizations lie in these outfits that are becoming more professional in their approach."

For the healthcare industry, the result of all of this is simple. Cybersecurity will continue to get more and more critical as attackers become more sophisticated, and the industry will need to pay up anyway, one way or another. It might as well make the investment up front in defensive software instead of waiting for a breach to happen and having to pay a ransom.

WHY CYBERSECURITY IS IMPORTANT

In case you're not convinced that cybersecurity is important, I thought I ought to share a few more examples. The healthcare industry has traditionally been pretty poor at securing its software and servers, but it's going to need to move with the times if it's to embrace the future. With increasing amounts of health data being stored online, the industry has a responsibility to patients to secure it.

Otherwise we see cases like the Bronx Lebanon Hospital Center breach in which medical records for 7,000 patients were compromised. The records "disclosed patients' mental health and medical diagnoses, HIV statuses and sexual assault and domestic violence reports" according to NBC News. It could be worse, though – one Canadian plastic surgery center and spa leaked before and after photos of patients' breast augmentations.[323] In October 2017, the BBC reported on a cyberattack on London Bridge Plastic Surgery (LBPS) in which the alleged hackers obtained photos showing various body parts of clients, including genitals.[324]

---

[323] See: http://bit.ly/canadadataleaks

[324] See: http://bit.ly/lbpsattack

It's interesting to note that the type of medical records that were exposed at Bronx Lebanon are protected under the Health Insurance Portability and Accountability Act (HIPAA). In its summary of HIPAA, the Department of Health and Human Services warned that "the rise in the adoption rate of [electronic health records] increases the potential security risks."[325]

Other breaches include research carried out by "white hat" hackers (i.e. those who wish to expose problems without causing harm so that vulnerabilities can be patched ahead of a more serious breach) from Independent Security Evaluators. They were able to gain access to patients' monitors, which "they could force to change at will – displaying false alarms or incorrect readings, which could easily lead to fatal treatment being given to patients." In their coverage of the research, Gizmodo added, "The prospect of a hack simply shutting down hospitals is scary enough on its own, but the paper demonstrates a malicious hacker could actively toy with equipment to kill patients. Equally bad are the flaws that enabled the hack: it's not one specific problem, but rather a systematic lack of good software and security policy that leave innumerable gaping holes."[326]

Even ancillary companies will need to take care. Doctors at Boston's Beth Israel Deaconess and the University of Pittsburgh Medical Center found themselves unable to use Nuance's eScription software after the company was hit by a cyberattack. As a result, they were forced to switch back to pen and paper, with the University of Pittsburgh explaining that its dictation and transcription services were offline with "no estimated time of resolution."[327]

And then there's the disturbing breach of the Heatmiser smart thermostat in which a hacker was able to remotely increase the

---

[325] See: http://bit.ly/patientrecordsbreach

[326] See: http://bit.ly/whitehatbreaches

[327] See: http://bit.ly/nuanceattack

temperature inside a house by 12ºC (22ºF).[328] This might not seem too relevant to healthcare, but if our health is in the hands of our smart devices then any breach of any part of the system is cause for concern.

EHR INTEROPERABILITY

We've already talked about how EHR interoperability is required for the future of healthcare to come about, and steps are already being taken to encourage it, although it's not necessarily the government that's leading the charge.

Healthcare IT News covered some interviews ahead of the Healthcare Information and Management Systems Society 2017 conference in which three major companies signaled a shift to more open EHRs.[329] Allscripts CEO Paul Black, Epic CEO Judy Faulkner and Cerner President Zane Burke explained that they're working on making their EHRs more open by "embracing APIs as a means to enable third-parties to write software and apps that run on their platforms."

Allscripts has already certified 5,000 developers to use their APIs, and over two billion API data exchanges have taken place on their platform since 2013. Epic is working on two new versions of its EHR and developing an API system called Kit that can access data from Caboodle, its data warehouse. Paul Black summarized the situation by explaining: "Instead of saying 'that was great, we're done' and sitting back in a rocking chair, now it's, 'Holy Moly, we have all of this data and what are we going to do with it? And how do we use all this data to drive more efficient, effective care that produces better outcomes for people who have serious issues?'"

Meanwhile, speculation is rife that Apple and Amazon are

---

[328] See: http://bit.ly/hackerthermostat

[329] See: http://bit.ly/moreopenehrs

getting ready to enter the EHR market, and while neither of them has officially revealed their intentions, we've already seen how they're making moves. Healthcare IT News has another fantastic article which summarizes just a few of the latest developments:[330]

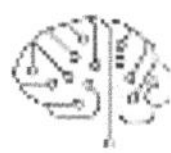

AETNA: The company is reportedly working with Apple to develop smartwatches that can monitor chronic diseases.

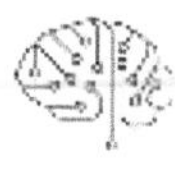

AMAZON PHARMACY: Some analysts believe that Amazon will move Echo/Alexa into the clinical realm and start releasing telemedicine and remote patient monitoring technology.

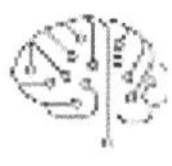

AMAZON'S SECRET LAB: Amazon has created a 'secret' lab at its Seattle headquarters to explore the healthcare sector, including the company's potential to disrupt EHRs and telemedicine.

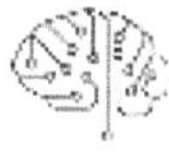

ARGONAUT PROJECT: Apple is working with this private sector initiative to integrate more electronic health data with the iPhone. It's also reportedly in talks with hospitals and other healthcare organizations to bring health records together on the iPhone.

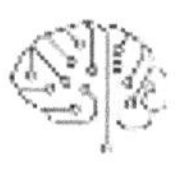

CAREKIT: Boston Children's Hospital has collaborated with Duke Health System to develop the new Caremap app using Apple's CareKit. It allows family members and caregivers to track and store health metrics and medical information so it can later be shared with doctors and clinicians.

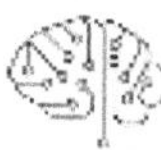

RESEARCHKIT: Launched two years ago, Apple's ResearchKit aims to use iPhones to gather health data so

---

330 See: http://bit.ly/appleandamazonehr

that researchers can tap into that information to conduct studies.

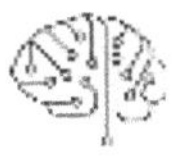

SUMBUL DESAI MD: Apple recently hit the news after hiring Sumbul Desai MD, who's also a clinical associate professor of medicine at Stanford and the vice chair of the Department of Medicine and chief digital officer at Stanford Center for Digital Health. She'll serve a senior role at Apple while continuing to see her patients at Stanford.

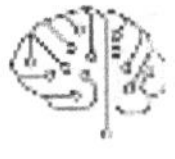

WEBMD: The popular health information site can now be accessed through Amazon's Alexa, meaning that people can ask questions about conditions, medication, tests and treatments right from their Echo, Echo Dot or Amazon Fire TV.

COMPANIES WITH TOO MUCH POWER

It's probably good news for us if Google, Apple and Amazon make moves on the healthcare industry, but we also need to make sure that we're not giving these companies too much power. Power corrupts; absolute power corrupts absolutely.

There are several reasons for this, the most obvious being that without constant competition, there's less incentive for innovation. But another big problem is that when one entity is catering to a huge proportion of the market, any decisions they make could have far-reaching consequences.

Apple recently refused to enable a feature called Advanced Mobile Location (AML) in iOS. As TheNextWeb explains, "Enabling AML would give emergency services extremely accurate locations of emergency calls made from iPhones, dramatically decreasing response time.[331] Google has already implemented AML on Android devices and the technology is saving lives, but Apple still refuses to use it."

According to TheNextWeb, "The majority of emergency calls today are made from cellphones, which has made location pinging increasingly more important for emergency services. There are many emergency apps and features in development, but AML's strength is that it doesn't require anything from the user – no downloads and no forethought. The process is completely automated. With AML, smartphones running supporting operating systems will recognize when emergency calls are being made and turn on GNSS (global navigation satellite system) and Wi-Fi. The phone then automatically sends an SMS to emergency services, detailing the location of the caller. AML is up to 4,000 times more accurate than the current systems – pinpointing phones down from an entire city to a room in an apartment."

There's no word as to whether Apple plans to implement the technology in the future, but we can assume from their silence that they probably don't. That's bad news for the 26% of Europeans who use an iPhone, but at least the 70.8% who use an Android device will be covered.

Meanwhile, the technology has already started to make a difference, as is the case with the eight-year-old Lithuanian boy who saved his father after finding him unconscious. The boy was unable to provide the address, but the inclusion of AML allowed emergency services to place the call within a radius of six meters (instead of 14 kilometers). It's estimated that if Apple added AML and the system was implemented everywhere in the EU, AML could save over 7,500 lives over the next ten years.

---

[331] See: http://bit.ly/appleaml

WHO WILL CALL OUT THE EMPEROR?

I recently came across an insightful article on LinkedIn in which the author pointed out that "truth in healthcare advertising and marketing would dictate that claims or advertising regarding doctors, hospitals, therapists, diagnostics, treatments and devices would discuss the price and quality of preventative, medical, surgical and palliative outcomes produced by doctors, hospitals, therapists or devices with their patients."[332]

He quite rightly explains that the current advertisements rarely make any claims about the quality or the price of the outcomes that they produce. But for the healthcare industry to work more efficiently, it needs to better understand the cost and quality of products and outcomes relative to other treatment options.

I'll let Dr. Green take it from here because I think he absolutely nails the situation in his article:

The insurance, pharmaceutical, EMR and pharmacy industries have been privy to this outcome data for decades, bartering, selling and trading private patient healthcare information for profit. Insurance actuaries don't like to guess. Wouldn't it be nice if patients and physicians [didn't] have to guess who the best doctors and hospitals [are]? The physicians who produce and manufacture clinical outcomes with patients have been prevented from access to this outcome data.

There's only one way to tabulate and reveal clinical outcomes and that is through the standardization and integration of interoperable electronic medical records (EMR/EHRs) and billing systems. The ability to digitally examine clinical outcomes according to all variables has existed for over two decades. Unfortunately, the insurance, pharmaceutical, [medical malpractice], [academic publishing], hospital and EMR/EHR and billing industries – together with their patron politicians – have prevented the establishment of interoperable EMRs due to the potentially devastating effects of revealed outcomes on those ancillary healthcare industries.

---

[332] See: http://bit.ly/howardgreenmd

Next time you witness marketing or see an advertisement relating to the doctors, hospitals, therapists or apps which manufacture preventative, medical, surgical and palliative outcomes with patients, ask yourself if there is any proof being presented that their outcomes are better than their competitors. Wouldn't everyone like to know what insurance companies have the best clinical outcomes and associated prices for their insured customers? Wouldn't everyone like to know what is the highest quality and lowest cost drug for which diseases and why? Wouldn't everyone like to improve under-performing doctors and hospitals via capitalistic market competition based on outcomes instead of simply suing his or her scrubs off?

Wouldn't everyone like to see advertisements in newspapers, billboards, in-flight magazines and on the internet for doctors, hospitals, therapists [and] pharmaceuticals which state that their quality and price is best for certain diseases and patients? Wouldn't everyone like to know if the VIP or concierge industry delivers improved outcomes justifying the costs? It's time to end empiricism in healthcare marketing, advertising and practice.

As Shakespeare's Hamlet said, "Outcomes are the thing wherein we'll catch the conscience of the king." Perhaps Hans Christian Andersen stated it best when he concluded in his short tale, "Without revealing clinical outcomes, the emperors of healthcare clearly have no clothes."

ANOTHER ELVIS MOMENT

Abhinav Shashank argues that healthcare needs another Elvis moment[333], pointing to when Elvis Presley accepted a polio shot on the Ed Sullivan Show, while the world – and the world's press – was watching. His action helped to raise immunization levels in the US from 0.6% to 80% within just six months.

The point here is that healthcare needs another Elvis moment if

---

[333] See: http://bit.ly/anotherelvis

it's to instill the future of healthcare in the hearts and minds of the general public. We live in a world of influencers, where marketing departments are approaching YouTube celebrities and trendy Instagrammers to spread their message. Unfortunately for medical practitioners and their patients, the healthcare industry is years behind the marketing industry in terms of its willingness to adopt new ideas and new technologies.

I'm not saying that the future of healthcare won't happen until IBM implants a smart device into Justin Bieber during the half-time break at the Superbowl – although that would be a good start. Instead, the healthcare industry will need to focus on improving its reputation at the same time as improving its usage of technology.

WHERE ARE THE WOMEN?

Healthcare has historically been somewhat male-dominated, particularly at the highest level. While I was working on the book, I received a frank email from a conference that was actively looking for female speakers because "if things [were] left to their natural course [they would] end up with a male-based agenda" which they believed "does not represent the future of our industry."

While it's great to see that such steps are being taken, it's also sad to see that they're necessary in the first place. Reducing the male bias could help to drive positive change for the industry – and it could also make absolutely no difference. The issue of gender disparity in healthcare is a sensitive and highly discussed issue, but it's a conversation that needs to be had. After all, in the future of healthcare, everyone needs to be represented.

It's something that we'll need to keep our eyes on as the healthcare industry develops. After all, technology is becoming more and more important to the field of healthcare, and the tech industry has historically been male-dominated. But let's hope that doesn't happen. In my view, the future of healthcare will need to be more feminine, at least in terms of its stereotypical traits. It needs to be caring, communicative and collaborative. There'll be no place for the 'masculine' traits of ruthlessness and determination in the pursuit of wealth at all costs. That's why it's inspiring to see amazing leaders such as NYC Health Business Leaders' Bunny Ellerin and Dr. Grace Cordovano of Enlightening Results ushering in this change.

REGULATORY CONCERNS

The slow adoption of tech in the healthcare space is due, in part, to the regulatory concerns and privacy issues that surround patient data. Justin Chadwick, director of analytics services and product marketing at Crossix Solutions, encapsulates the issue by explaining, "Opportunities are emerging and leveraging data as a way to join the dialogue with patients is key. The industry's digital paralysis has largely stemmed from a general lack of understanding of and comfort with using data in the privacy-concerned world."[334]

Gavin Johnston, group planning director of Intouch Solutions, believes that the industry's conservative approach is understandable but argues that a high concentration of patients are becoming involved in their own treatment through digital technology and that the healthcare industry is being forced to act. "Going forward we are going to see more risk taking and innovation," he says. "And whichever company leads this movement will dominate any market it is in and have the most positive effects for patients."

Meanwhile, Liberate Health president Richard Nordstrom argues, "Digital is disrupting every aspect of the practice of medicine, empowering clinicians and patients alike. Who better to help provide these meaningful services than the pharma industry, [which] understands clinician workflow and patient engagement and has the marketing expertise and resources?"

---

[334] See: http://bit.ly/digitalparalysis

For me, the whole debate around regulation and the interoperability of data comes back to Joe Biden's exchange with Judy Faulkner from Epic. Patients have a right to their data – and it's up to them to decide who they want to share it with.

INTEROPERABILITY

Interoperability is essential for the future of healthcare to be brought about. By aggregating information from across the community in different systems, the data can be parsed so that we can deliver improved care and a variety of new tools to better manage both individual patients and the population as a whole.

On top of that, as we inevitably move closer and closer towards a value-based system in which we pay for the quality of the care instead of the quantity, interoperability for EHRs becomes essential. We need to be able to analyze performance data openly for different medications and treatment options so that we have the information we need to make value-based decisions.

So what's stopping us? Well for starters, the government could do more to push for this change to happen, and it's also true that some providers don't actually want to share information. After all, many companies are benefitting from the current system and it can be tempting to see data as just an asset to be guarded instead of as something to be shared around for the greater good.

In the current fee-for-service world, there's neither a need nor an incentive to share patient data. But as we move closer to the future of healthcare, interoperable data will become essential. It comes down to the fact that without access to comprehensive amounts of data on patients, providers simply can't provide the best possible value-based care for their patients.

But how can we create interoperable EHR systems? In an article he shared on LinkedIn, Abhinav Shashank suggested that the answer to the problem could be application programming interfaces (APIs).[335] These are essentially little pieces of connecting technology which allow you to integrate one service with another. When you log into a website using your Facebook account, it uses Facebook's API to authenticate you.

According to Shashank, using APIs has numerous benefits including real-time access to data and time and resource savings due to autonomous and easy-to-manage processes. Best of all, just about anyone can release an API – and just about anyone can take an API and build on it to make a better system. Everybody wins.

CONSOLIDATION

One of the big trends that we're likely to see in the coming years is the increasing consolidation of the healthcare industry. Huge companies will buy the smaller innovators in an attempt to stay on top of the market, and we're already seeing that in action thanks to high-profile acquisitions like KKR's purchase of WebMD for $2.8 billion. This will fit beside "a large portfolio of B2B and B2C websites in verticals like automotive, health, home, travel and legal" under KKR's Internet Brands umbrella.[336]

WebMD is undeniably a popular website. It was founded in 1996 and went on to become the go-to source for medical information for an entire generation. It's the Wikipedia of health and ranks 36th in comScore's list of the top 50 websites in the US, attracting 71.7 million unique visitors in June 2017 alone. It was the only site in the medical sector to make the list.

Bob Brisco, the CEO of Internet Brands, released a statement

---

335 See: http://bit.ly/interoperabilityehrs

336 See: http://bit.ly/kkracquisition

saying: "Since its founding, WebMD has established itself as a trusted resource for health information. We look forward to delivering that resource to even more users by leveraging our combined resources and presence in online healthcare to catalyze WebMD's future growth." Meanwhile, WebMD CEO Steven L. Zatz MD said, "We believe that this transaction will provide additional flexibility and resources to deliver increased value to consumers, healthcare professionals, employers and health plan participants. I am confident this will be an exciting new chapter for WebMD."

As well as WebMD, Internet Brands owns DentalPlans.com, eHealthForum.com, HealthBoards.com, VeinDirectory.org and FitDay.com, while last year it acquired healthcare-focused marketing automation company Demandforce from Intuit.

But is this consolidation a good thing? Perhaps. It's still in its early days and so we can only speculate, but it's a pretty safe bet that it will improve overall access to information. It will also create increased competition as new startups try to take advantage of the healthcare boom, and those same startups will have access to the financial backing of established behemoths if they're acquired. It's a win/win situation for everyone – in theory, at least.

EMERGING MARKETS

One of the challenges that global healthcare will face is that of emerging markets, which typically have less access to technology. They certainly won't be able to spend as much on healthcare as we do here in America.

But the lack of budget is just one of many challenges that emerging markets will face. Eric Bellman, deputy bureau chief for South Asia at the Wall Street Journal, penned an article arguing,

"Tech tricks are no fix for developing-world problems."[337] He's half-right, but he's unaware of the full story.

It's true that fancy machines and supercomputer algorithms are prohibitively expensive in developing countries, but as far back as 2008, Madeline Drexler was showcasing tech's potential in an article for the New York Times about a "sturdy, low-cost incubator, designed to keep vulnerable newborns warm during the first fragile days of life."[338]

You're probably thinking that an incubator doesn't sound revolutionary, but that's because you don't know what it's made from. As Drexler explains, "The heat source is a pair of headlights. A car door alarm signals emergencies. An auto filter and fan provide climate control." This cuts the cost from over $40,000 for a high-end American model to less than $1,000, but the real benefit comes from the fact that it's made from recycled car parts. The more expensive machines are complicated and need specialist repairmen when something goes wrong, something that most developing countries don't have access to. But when the incubators are made from recycled car parts, any automotive mechanic can carry out the repairs.

It's a simple idea, but it's also effective. According to Dr. Kristian Olson, the principal investigator on the project, the main causes of newborn death – infections, preterm birth and asphyxiation – are readily treatable with the right expertise and equipment. He called them the "low-hanging" fruit of global health interventions, adding, "It's so frustrating to see these preventable deaths. They won't name babies in Aceh, Indonesia, until they're two months old. It's a cultural adaptation to expect a death."

The moral of the story here is that the developed world approach to healthcare won't work in developing markets. Instead, entrepreneurs and innovators will need to think of entirely new ways to cater to these markets, focusing on getting the most value

---

[337] See: http://bit.ly/techtricksarenofix

[338] See: http://bit.ly/drexlerincubator

from the lowest cost.

It's no different to the value-based healthcare that this book is helping to usher in. After all, the future of healthcare is personalized medicine. If medicine should change from one person to another, what does that say about the different approaches that will be needed in developing and developed countries?

CIVIL RIGHTS

One of the more controversial aspects of the future of healthcare is its impact on civil rights. Let's say that a doctor has told an alcoholic that if they don't stop drinking, they'll die. What if their Alexa device literally locked them into their house if they thought they were at risk of relapsing? Strictly speaking, it's in the patient's best interests, but at the same time it's essentially false imprisonment. And even if it's judged to be false imprisonment in a court of law, who's guilty of the crime? Will the prisons of the future be filled with Amazon Echoes?

Of course, there are less extreme quandaries for us to solve. Let's say that someone's trying to quit smoking and their Alexa tells their doctor every time it detects cigarette smoke. What if the patient explicitly commands the device not to tell the doctor? Who does it obey?

We're already starting to see these issues manifesting, and we'll only see more of them in the future. A great example from the present is explained by Grace Ballenger in an article for Slate.com, where she talks about a groundbreaking ruling in Ohio where a judge "ruled that a man's pacemaker data could be used against him in a criminal case, opening the door for a new type of electronic surveillance."

But this raises all sorts of questions about our privacy, and while some would say that they have nothing to hide, I beg to differ. There was a fantastic section in Oliver Stone's 2016 biopic *Snowden*, which followed the story of the notorious NSA whistleblower, where he tells his girlfriend that Russian hackers could access the webcam of her laptop. She replies, "Whatever, it's not a big deal. You shouldn't let it bother you. I'm not hiding anything." Snowden replies, "That's such a bullshit line. Everyone does." When his girlfriend protests that she doesn't, he says, "The other day your computer was open and I happened to notice you were on the site where we met, and you were looking at other guys."[339] Predictably, she then gets upset because her privacy was breached and an argument ensues, but it does go to show that we all have something to hide. Another example of how we take our privacy for granted comes via Buzzfeed's "Hardest Game of Would You Rather": "Would you rather have your entire Google search history sent to your parents or have all the photos on your phone sent to your parents?"[340] Perhaps unsurprisingly, 75% of people would rather send their parents the photos.

The civil rights issue is a valid concern and something that will need to be addressed, which is why regulators need to be involved early on to set privacy guidelines. My personal point of view is that as long as the health issue only affects the patient and doesn't put any other person at risk, the patient's decision should always be final. The power should rest with the patient, whether they're filing a do not resuscitate (DNR) request or whether they're refusing to allow medical practitioners to access their health data.

But let's face it – we're usually willing to give up our privacy in exchange for the greater good. As a society, we're more than happy to share personal information on Facebook and Twitter, so we'd be crazy not to share personal information with the doctors and institutions who could one day save our lives.

---

[339] See: http://bit.ly/Snowdenscript

[340] See: http://bit.ly/hardestgamewyr

ASIMOV'S LAWS AND THE UNCANNY VALLEY

Throughout this book, one of the big problems that we've discussed is the role of AI when human lives are at stake. If there's a situation in which a patient needs an operation with a success rate of 80%, should a clinical decision support tool recommend it even though there's a 1-in-5 chance that it will lead to the patient's death?

This is a classic case of the prisoner's dilemma. One solution to this could be to apply Isaac Asimov's three laws of robotics, which the science fiction author introduced back in 1942 as the rules that govern the behavior of robots:[341]

1. A robot may not injure a human being or, through inaction, allow a human being to come to harm.

2. A robot must obey the orders given to it by human beings, except where such orders would conflict with the First Law.

3. A robot must protect its own existence as long as such protection does not conflict with the First or Second Law.

It's interesting to note that Asimov later added a fourth law that outranked the others: "A robot may not harm humanity, or, by inaction, allow humanity to come to harm." The first three laws would cover how robots and AI would treat individual patients. The fourth would cover how they'd treat the global population as a whole. But there's a lot of work to be done to determine what this would look like in practice – and whether Asimov's laws are even the solution.

Asimov also coined the term "Frankenstein complex" for the fear of mechanical men, a term which reminds me of Masahiro Mori's "uncanny valley". But Mori's theory is arguably more in-depth while

---

[341] See: http://bit.ly/timetravelbots

offering a solution of sorts.

In an article for the New Scientist, Laura Spinney described the uncanny valley hypothesis as "the notion that the more human-like a non-human character becomes, the more we like it – until suddenly, we don't."[342] Human beings naturally react better to robots if they look and act more human, but there's a point at which they become too human and we start to get freaked out. On some basic level, we know that something's wrong, and the idea of a machine masquerading as a man can make us very uncomfortable indeed.

This is a stark warning for developers of AI and virtual assistants. While these new technologies will be vital if the future of healthcare is to come about, they'll also need to be careful not to be too creepy. They need to be realistic, but not *too* realistic.

KEEPING TRACK OF PATIENTS

Every year in the US, 45 million people move, three million change marital status and 21 million change employment status.[343] This makes it very difficult to keep an up-to-date database, especially without interoperable EHRs which could spot mismatched data and automatically update all records to show the latest, most accurate information.

One solution to this would be to encourage people to take a greater role in their healthcare by giving them access to their medical records and encouraging them to ensure that they're up to date. But this might also require a groundswell amongst the general public, especially in an age in which people think they need to be sensitive about their personal information.

Personally, I don't understand it. Why are we so sensitive? We share our information with Facebook, Google and Amazon, and

---

[342] See: http://bit.ly/exploringuncanneyvalley

[343] See: http://bit.ly/socialdeterminantshealthcare

when we share data with these business entities, we're doing it solely to have access to the service. When it comes to healthcare, we're talking about sharing data so it can save lives – both your own life and the lives of other people through the lessons that can be learned from the aggregation of global health data. Would you rather share your data with Facebook to get targeted ads or with researchers who could use that data to improve the quality of life for you, your loved ones and for generations to come? These are some of the questions we need to ask ourselves.

NEW PARTNERSHIPS AND MISSED APPOINTMENTS

One exciting area of potential growth is that of value-led partnerships. We can already see this in action when we look at how tech companies are getting into the game, such as the partnership between IBM's Watson and the NHS.

One of my favorite examples of these partnerships comes via the CareMore Health System, "a healthcare delivery system that treats patients with the care and dignity they deserve." The company has partnered with Lyft to "provide non-emergency medical transportation for Medicare Advantage and duals beneficiaries."[344] The data shows that patients using the service wait less than nine minutes to be taken to their appointment, which is a 30% reduction in wait times. Meanwhile, costs per ride have dropped from $31 to $21, and the partnership as a whole is supremely useful for the people who need help the most.

In fact, it's estimated that 3.6 million Americans either miss or delay the receipt of non-emergency care every single year, purely due to problems with transportation. Worse, it's often those with chronic diseases or with serious health conditions that face the greatest barriers to seeking treatment.

---

[344] See: http://bit.ly/caremoresystem

It'll be interesting to see the ways in which technology and healthcare companies are able to work with each other. I'll also be keeping my eyes peeled on the issue of medical transportation for those who find it difficult to get to and from their appointments. Self-driving ambulance, anyone?

WHY IT'S SO HARD TO CRACK HEALTHCARE

Thomas Goetz, the CEO and co-founder of Iodine, published a fascinating piece on Inc.com where he talked about how he tried – and failed – to revolutionize healthcare.[345] When Iodine launched in 2013, Goetz says it was "part of an epidemic of startups seeking to change healthcare." He also says that there was "great enthusiasm on the part of entrepreneurs and investors alike" for healthcare disruption, explaining that there was over $17.8 billion raised between 2011 and 2016 for healthcare startups.

But investors only invest if they think they're going to get a return, and Goetz says "the return on that investment has been dismal at best." He says the disruption everyone was waiting for never happened – and that it probably never will, although I disagree as I'm optimistic – and says that the reason is pretty simple. "Healthcare is different," he explains. "It's highly regulated, which makes rapid transformation difficult. The incumbents are massive enterprises with multiple services, so challenging them is nearly impossible. It isn't a market-driven industry that responds to better, cheaper, faster. You can't price-shop. The government is the biggest customer. All the incentives are misaligned."

So that's why it's so hard to crack healthcare – there isn't just one challenge, there are many, and most of them are deep-seated and play into the hands of the established companies.

Luckily, there's an answer, and Bayer's global head of digital

---

[345] See: http://bit.ly/goetzsohardtocrack

health incubation and innovation Eugene Borukhovich is only too happy to share it. "The ONLY key to innovation is people," he says. "Brilliant, driven, passionate people who are willing to put politics aside and focus completely and solely on us, human beings. People can fix this, but only people who are open to change, who think exponentially, who are open to experimenting and learning, who are open to intense challenges ahead to push the status quo in order to save or extend a life even if it's only a day, a month, a year at a time."[346]

So the key to innovation is people. People like you and I.

---

[346] See: http://bit.ly/thekeytoinnovation

CHAPTER SIXTEEN: TO THE FUTURE!

"YOU NEVER CHANGE THINGS BY FIGHTING THE EXISTING REALITY. TO CHANGE SOMETHING, BUILD A NEW MODEL THAT MAKES THE EXISTING MODEL OBSOLETE."

- BUCKMINSTER FULLER

NEW TECHNOLOGIES have unimaginable potential when it comes to revolutionizing healthcare. I foresee a future within our lifetimes in which humans and machines are able to truly partner together for the best possible outcomes.

I agree with Klaus Schwab, the founder and executive chairman of the World Economic Forum. He believes that we're entering a fourth industrial revolution[347] in which new technologies have the potential to be the biggest thing to hit healthcare since the discovery of penicillin. People like John Brownstein are already making waves by studying digital data and carrying out research to find out whether Yelp reviews could be an indicator of food poisoning outbreaks or how public conversations on Twitter could highlight public health issues like chronic disease and gun violence.[348]

Ultimately, in my vision of the ideal future, real world data will be captured about every patient through data sources like

---

[347] See: http://bit.ly/airedesignhealth

[348] See: http://bit.ly/harvardjohn

Ancestry.com, 23andMe, wearable devices and smartphones, and AI will gather and analyze this data and then project it onto a screen in the doctor's office. That way, when a patient visits a physician, they can make a holistic and objective decision instead of basing them solely on subjective information that's presented by the patient and subjectively interpreted at the point of care.

Eric Topol explained in a recent article for The Wall Street Journal that "radical new possibilities in medical care are not some far-off fantasy." He says, "Last week in my clinic I saw a 59-year-old man with hypertension, high cholesterol and intermittent atrial fibrillation (a heart rhythm disturbance). Before our visit, he had sent me a screenshot graph of over 100 blood pressure readings that he had taken in recent weeks with his smartphone-connected wristband. He had noticed some spikes in his evening blood pressure, and we had already changed the dose and timing of his medication. The spikes were now nicely controlled. Having lost 15 pounds in the past four months, he had also been pleased to see that he was having far fewer atrial fibrillation episodes – which he knew from the credit-card-size electrocardiogram sensor attached to his smartphone."[349] And this is just a glimpse into the potential of the new technologies on the horizon.

In this chapter, the final chapter of the book, we'll take one last look at the technologies that we're likely to see in the future. Let's go.

ENCODED DNA

It's impossible to understate just how quickly technology is evolving. New developments are constantly shaping the world around us, as you're about to find out.

Let's take a look at E. coli bacteria, which lives in the intestines of

---

[349] See: http://bit.ly/smartmedicinesolution

living creatures and can cause infection and gastrointestinal issues. E. coli is notable for its easy-to-edit genome, which has been manipulated by scientists to include a GIF image of a galloping horse that they were able to retrieve and reconstruct with 90% accuracy, even after multiple generations of bacterial growth.[350]

Scientists have already been able to encode all 587,287 words of War and Peace – as well as an OK Go music video – within DNA. But the E. coli case is interesting because it's the first time that data has been encoded within living DNA, and while a horse GIF might not seem medically relevant, it opens up all sorts of potential for the future. The most interesting possibility is the way that living cells might be able to store a recorded history, enabling specialist healthcare practitioners to look at exactly what's happening to cells within the human body.

Scientists at Microsoft, meanwhile, are investigating the potential of using DNA as a storage device. The costs need to come down by a factor of 10,000 before DNA storage can compete with traditional storage methods, but it's only a matter of time. Better still, once DNA storage becomes mainstream, it could vastly reduce the amount of energy that we expend on server racks and electronic apparatus – and it could keep data intact for over 100,000 years.

But most importantly of all is the potential for this type of technology to fight diseases. As Vox explains, "In a paper published in the prestigious journal Nature, a team led by Shoukhrat Mitalipov of Oregon Health and Science University described how it used CRISPR/Cas9 to correct a genetic mutation that's linked to a heart disorder called hypertrophic cardiomyopathy in human embryos. And they did it all without the errors that have plagued previous attempts to edit human embryos with CRISPR."[351]

This opens up an interesting debate. Studies like this show how we could fight diseases at a genetic level, which could pave the way for a whole new generation of preventative medicine. It could "one

---

[350] See: http://bit.ly/crisprhorse

[351] See: http://bit.ly/crisprexperiment

day mean the ability to create smarter or more athletic humans" – or so-called "designer babies". It's an ethical minefield, but that's to be expected.

It reminds me of that quote from Spiderman: "With great power comes great responsibility." It's up to us as a race to make sure that we don't abuse it.

FUTURE TECHNOLOGIES

We've covered all sorts of different technologies throughout this book, and the likelihood is that the future of healthcare will see all of them brought together to form a cohesive whole. Still, I thought I'd take the opportunity to share just a few more of the up-and-coming technologies that have the potential to change healthcare for the better.

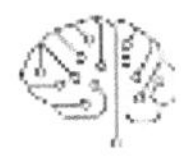

AUTOMOTIVE HEALTH TECH: Hyperadvancer founder Bart Collet shared a post on LinkedIn where he talked about how the ability to monitor our health while driving will soon be an essential component of our cars. He says, "Are you almost falling asleep? Is your [heart rate variability] predicting a [sudden cardiac death]? Are your pupils almost entirely black from cocaine abuse? In each of these cases the car should stop. But this type of 'safety monitoring' only scratches the surface of possibilities."[352]

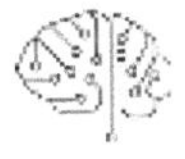

SELFIE APPS: The BBC recently reported on the BiliScreen app, an app that aims to make it easier to detect the early signs of pancreatic cancer through a simple selfie. It was developed by a team of medical

---

[352] See: http://bit.ly/healthcartech

clinicians and computer scientists from the University of Washington.[353]

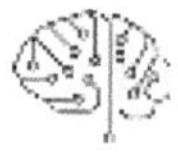

3D Printed Heart: While it might just be a prototype, Swiss researchers at ETH Zurich successfully 3D printed a heart that pumps blood and fluid just like a real heart. The idea is that it will one day replace blood pumps, which can cause complications when the patient lacks a pulse, although it currently only lasts for around 3,000 beats – or 30-45 minutes.[354]

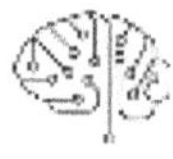

AUTOMATED CPR: The innovative AutoPulse by Zoll is a customizable and automated CPR device that claims to automatically calculate the size, shape and resistance of patients' chests to deliver unique compressions for each patient. Perhaps more importantly, it could free up paramedics' time in emergency situations and allow them to carry out other life-saving work.[355]

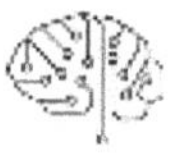

SMART CLOTHING: Researchers at the Georgia Institute of Technology have published a paper based on their experiences developing smart clothing technology that could capture patients' data to create and update medical records.[356]

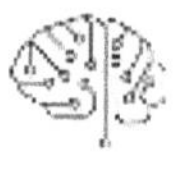

SMART CROSS-TRAINER: Engineering student Ronan Byrne created an innovative device called the Cyxflix, "an exercise-powered home entertainment system that takes the guilt out of binge-watching your favorite shows by working your ass off for the right to

---

[353] See: http://bit.ly/selfieappcancer
[354] See: http://bit.ly/3dprintedheartbeat
[355] See: http://bit.ly/automatedcprdevice
[356] See: http://bit.ly/smartclothingpaper

see them." If you stop exercising, the show cuts out.[357]

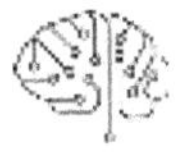

SELF-DRIVING SIGN LANGUAGE BUS: This one's probably my favorite of them all. Technology Review reports that Local Motors and IBM are working on "an autonomous electric shuttle bus with technology that assists people with a range of disabilities." In her article for the site, Elizabeth Woyke said, "For deaf people, the buses could employ machine vision and augmented reality to read and speak sign language via onboard screens or passengers' smartphones."[358]

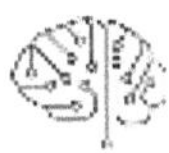

AI FOR CONTINOUS PATIENT MONITORING: Adam Hanina and his amazing team at AiCure have likened their technology to a personal trainer in a gym working directly with a client to achieve their goals. It involves facial recognition and motion-sensors in a mobile device and records patients taking their medication. Then it transmits that data back to a clinician through a HIPAA-compliant secure network so that the clinician can confirm that patients are taking their medication. It also has the potential to flag adverse events or potential barriers and works with patients to overcome them. I was fortunate enough to try out this technology first-hand and I must say I was completely blown away. The darn thing works great. All my attempts to trick the technology were in vain.

---

[357] See: http://bit.ly/netflixcrosstrainer

[358] See: http://bit.ly/selfdrivingbus

NANOTECHNOLOGY

As an external advisory board member on nanotechnology for the Massachusetts Institute of Technology's MIT.nano project, I have no doubt that this technology will transform the field of medicine as we know it. It's no secret that technology platforms are getting smaller and smaller. Nanotechnology is just the natural next step. We're living at the dawn of the nanomedicine age in which nanoparticles and nanodevices will soon operate as precise drug delivery systems, cancer treatment tools and tiny surgeons.

Nanotechnology might sound like the stuff of science fiction novels, and in many ways it is. But it's also a rapidly growing field for medical research and development, and there are plenty of real-world possibilities when it comes to deploying nanotechnology. Diabetes is a condition often caused by low levels of insulin or the absence of insulin as a result of its destruction by immune cells. Insulin functions as a vehicle that transports sugar from our blood to other cells for processing and eventual energy production. Imagine a nanometer sized cage that lets out insulin but doesn't get attacked by our immune system. How about a nanorobot delivering dopamine directly to the brain for treating Alzheimer's? And how about stealthily injecting chemotherapy into cancer cells while keeping healthy cells untouched? Use your mind's eye for a second and imagine microscopic robots patrolling your entire body and sending alerts to your smartphone that a disease is about to develop in your body. These scenarios exemplify precision medicine at its best.

Of course, it's not necessarily easy to work with nanotechnology. Nanoscientists work at incredibly small scales. A nanometer is a billionth of a meter. But one billionth of a meter might be as abstract to many as is a trillion-dollar national deficit. To put the miniscule scale of this type of technology into perspective, a human hair is around 75 microns or 75,000nm (nanometers) in diameter. The relationship between a nanometer and that hair is similar to the relationship between one mile and an inch: one mile is 63,360 inches. A human red blood cell is 6,000-8,000nm across, and the Ebola virus is about 1,500nm long and 50nm wide.

DNA-based origami robots, ant-like nanoengines and sperm-inspired microbots are just a few in a long list of new technologies destined to enhance the future of healthcare. Let's take a short dive into three main ways this army of nanorobots can be put to use:

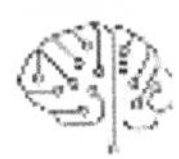

MORE ACCURATE DRUG DELIVERY: We touched on this earlier when we talked about nanotechnology's potential for targeting only the cells which are cancerous. Traditional technology can't differentiate between healthy and diseased tissues. Nanotechnology can.

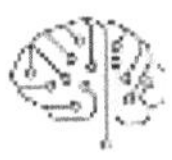

MORE EFFECTIVE CANCER TREATMENT: Nanotechnology isn't just for treating cancer patients, but it *is* one of the areas with the greatest potential for improved outcomes. Radiation and chemotherapy are currently our most effective treatment options, but nanotechnology could blow them out of the water.

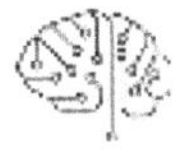

NANOPARTICLES AS INFORMATION TOOLS: Essentially, nanotechnology could allow us to constantly receive diagnostic information and therefore to target diseases before they become a problem. Researchers at John Hopkins University have already developed robots that are 1 millimeter across and which can take biopsies from inside the colon – and this is just the beginning.

TRANSPARENT PRICING

The issue of pricing has always led to debate in the healthcare industry. In an article for Forbes, venture capitalist Dave Chase tackles the issue of pricing failure – i.e. when there's no correlation between price and quality – explaining that it's "pervasive in healthcare."[359] Luckily, there are alternatives. Chase cites the case of Ascension's St. John's Health System, the largest nonprofit health system in America, which is joining the transparent medical network (TMN).

TMN is all about allowing patients and their doctors to "see the true costs of [healthcare] in advance in plain English", and the trend has been found to reduce surgical costs 50-90% versus the norm.[360] David Pynn, CEO of Ascension St. John, explained, "I believe that health systems like St. John need to do whatever they can to lower the costs of healthcare, and I think that population health is a reality that is coming. I don't think higher deductibles and shifting costs to employees is the answer."

Chase explains, "Pynn highlights the blunt-instrument tactic (high deductibles) that many employers have resorted to out of frustration with out-of-control costs. This pioneering hospital had the benefit of seeing forward-looking employers in their community successfully slay the healthcare cost beast while providing great healthcare benefits. As this small, Tulsa-based manufacturer demonstrated, it's possible to lower per-capita health benefits by 30% while improving benefits for their employees. Further, it allowed the manufacturer to invest money that would have otherwise been squandered on unnecessary healthcare costs into R&D – more than doubling what competitors spend on R&D."

Perhaps the most interesting thing of all about the adoption of transparent pricing is that patients often "self-deny" a claim because there are better options available or because they simply don't see the value of the treatment. As Chase explains: "A well-priced but unnecessary procedure is nearly as bad as an unnecessary procedure made worse by pricing failure."

Cutting down on wastage is important of course, but as Haider Warraich explains in a piece for Stat, "While several measures have been developed to curb wastefulness, some of these have had the

---

[359] See: http://bit.ly/tmnnonprofit

[360] See: http://bit.ly/healthreformcutscosts

unintended consequence of reducing access to potentially lifesaving treatments or therapies." He also points out that African-Americans are especially prone to underusing medications due to cost. "In other words," Warraich says, "well-meaning policies meant to curb excess can limit access."[361]

The result of this is at odds with the future of healthcare, which is all about preventative care. Warraich explains, "Medical interventions aimed at preventing disease or keeping it from escalating are universally compared to those with shorter-term effects. Consider people with asthma. It's recommended that they use a daily "controller" inhaler to ease the airway inflammation at the root of the disease, as well as a "rescue" inhaler that provides only short-term relief when symptoms appear. Yet controller inhalers are widely underused while many individuals with asthma overdose on rescue inhalers."

MICROCHIPPING

Will we all wear microchips in the future? A growing number of people believe that the answer to that question is 'yes'.

It makes sense. After all, it would allow us all to carry our entire medical history around with us so that if there's an accident, paramedics could easily scan the chip and check their patient's identity, allergies and supplementary information whether they're conscious or not. It would also reduce the need for identifying people through DNA and dental records.

Of course, not all people agree with microchipping. A great example of this was covered by CNBC in an article about a Wisconsin firm that embedded microchips in their employees "to ditch company badges and corporate logons". The article explained, "Religious activists are so appalled, they've been penning nasty 1-

---

[361] See: http://bit.ly/evidencebasedhealthcare

star reviews of the company, Three Square Market, on Google, Glassdoor and social media."[362]

The article also suggests that "in the future, consumers could zip through airport scanners sans passport or driver's license, open doors, start cars and operate home automation systems. All of it, if the technology pans out, with the simple wave of a hand." It also points out that the embedded chip isn't a GPS tracker, which means that it can't track employees' movements as critics initially feared, but there's no reason why that technology couldn't be added in the future. And if we all had one of these chips, the benefits to law enforcement would go above and beyond identifying John Does. They'd be able to track down missing persons and analyze the last minutes of murder victims – and their suspects.

Three Square Market is far from the only company using microchipping technology. The Los Angeles Times reported on the case of a Swedish start-up hub called Epicenter which "offers to implant its workers and start-up members with microchips the size of grains of rice that function as swipe cards to open doors, operate printers or buy smoothies with a wave of the hand." Ben Libberton, a microbiologist at Stockholm's Karolinska Institute, said, "The data that you could possibly get from a chip that is embedded in your body is a lot different from the data you can get from a smartphone. Conceptually, you could get data about your health, you could get data about your whereabouts, how often you're working, how long you're working, if you're taking toilet breaks and things like that."[363] And yes, this does all sound scary and Orwellian. We already live in a society in which closed-circuit television cameras are always watching our every move. We're also already slaves to the data that we create. As we start creating large amounts of data about our own bodies, there's potential for that data to be abused. Imagine a company disciplining an employee because their implanted GPS tracker shows that they're taking too many bathroom breaks. Now

---

[362] See: http://bit.ly/wisconsinmicrochips

[363] See: http://bit.ly/microchippingarticle

imagine them docking someone's wages because they're not typing fast enough. Or a totalitarian government knowing where all of its citizens are at all times.

Amal Graafstra, the founder of online biohacking store Dangerous Things and a double RFID chip implantee, believes that "the time is now" for chips to become more mainstream. He says, "We're going to start to see chip implants get the same realm of acceptance as piercings and tattoos do now." Call me a naïve futurist, but this makes sense to me, and it may be the best way for microchipping technology to develop. It removes many of the barriers and complaints that people have by relying on people who actively choose to be microchipped, and in the same way that some people hate tattoos and other people cover their entire body, some people will adopt new technologies and others will wait.

One final thing to mention here is the potentially prescient words of Noelle Chesley, an associate professor of sociology at the University of Wisconsin-Milwaukee. Chesley says, "It will happen to everybody. But not this year, and not in 2018. Maybe not my generation, but certainly that of my kids." She adds that corporations are slower to adapt to changes like these and that younger employees will naturally be more open to microchipping than older ones. She believes that consumers will adopt microchipping before companies do, adding, "Most employers who have inter-generational workforces [will] phase it in slowly. I can't imagine people my age and older being enthusiastic about having devices put into their bodies."

I can certainly envision future generations signing up to be chipped. But don't take my word for it. Just wait and see.

HEALTH SCORES

Imagine a world where many of your daily activities were constantly monitored and evaluated. What you buy online and in stores. Where you are at any given time. Who your friends are and how you interact with them. What you like and don't like. How many hours you spend on your mobile device, playing video games or watching content. Your genetic profile, your origin and your family history. Where you work, how much you earn, what your favorite restaurants are and what bills and taxes you pay.

It's not hard to picture because most of that already happens thanks to all those data-collecting goliaths like Google, Instagram, Facebook and LinkedIn, as well as smaller companies like Ancestry and 23andMe or health-tracking providers such as Fitbit and Apple. But now imagine a system where all these behaviors are rated as either positive or negative and distilled into a single number, according to rules set by the health authorities. That would create your health score, which would be publicly ranked against that of the entire population and used to determine your eligibility for health insurance, life insurance or a job, as well as what you should eat or even who you should marry.

Is this a futuristic vision of an Orwellian Big Brother out of control? No. According to an article recently published in Wired Magazine, it's already underway in China, where the government is developing the Social Credit System or Citizen Score to rate the trustworthiness of its 1.3 billion citizens.[364] For now, participating in China's Citizen Score is technically voluntary, but by 2020 it will be mandatory. The behavior of every single citizen and legal "person" (which includes every company or other entity) in China will be rated and ranked, whether they like it or not.

In the future, everyone will have a health score that's similar to the credit score that's used by banks for giving out loans. That way, the score can be used by life insurance and health insurance companies to determine our premiums, and as wearable health trackers become commonplace, sleep and physical activity can be tracked over time as a general metric to determine how healthy our lifestyle is. It will make the intangible tangible.

Connected devices and the internet of things will monitor our activities and upload that data. This will be factored into an

---

[364] See: http://bit.ly/chinesetrustscore

algorithm to generate an overall score, which can increase or decrease in real-time. People will be able to see their overall fitness going up and down as they're working out at the gym or eating takeaway pizza and watching Netflix.

This concept might sound like 1984 meets Pavlov's dogs: Act like a good and healthy citizen, be rewarded and be made to think you're having fun. It's worth remembering, however, that personal scoring systems have been present in human societies for decades. The notion of a health score is inevitable in the future of healthcare.

THE LIVES OF OUR GRANDCHILDREN

All of the changes that we've discussed so far will all come together to create a healthcare system that's virtually unrecognizable. The lives of our grandchildren will be completely different to the lives we live today and any would-be time traveler would be amazed. This is known as the singularity – the point at which civilization will change so much that its technologies are incomprehensible to previous generations. And like the singularity of a black hole, it's a point of no return. There'll be no turning back – but luckily, we won't want to.

According to Gizmodo, Longevity scientist Aubrey De Grey believes that "extreme human longevity (in the hundreds of years) is a crucial part of the singularity."[365] He suggests that a child born in 2010 could live to the age of 150 – or even older. Perhaps that's true, but I'd add a word of caution. The future of healthcare isn't all about longevity. Who wants to live to a thousand if every day is torture?

Meanwhile, artificial intelligence researcher and *Flesh and Machines* author Rodney Brooks says that as much as we might like to imagine it, it's "unlikely that [we'll] be able to simply download our brains into a computer anytime soon." However, he adds that

---

[365] See: http://bit.ly/gizmododegrey

"the lives of our grandchildren and great-grandchildren will be as unrecognizable to us as our use of information technology in all its forms would be incomprehensible to someone from the dawn of the twentieth century."

By 2030, it's estimated that one in five Americans will be over the age of 65, and by 2056 there will be more US seniors than people below the age of eighteen.[366] Meanwhile, according to a survey from the American Association of Retired Persons (AARP), 87% of people over the age of 65 would prefer to stay at home or in their communities as they get older.[367] Fortunately, as the future of healthcare is ushered in, our grandchildren will stay healthier and more active for longer, making it possible for them to maintain their independence. And as the size of this elderly population continues to grow, we can expect manufacturers and developers to continue to develop new technological applications to cater to them.

GAMIFICATION

Gamification is the process of adding the functionality of traditional games into other contexts. An example of this could be a leaderboard for a sales team or the badges that you receive on TripAdvisor for reviewing attractions. They're used to incentivize users to take action, and so it's no surprise that there's growing interest in gamification from the healthcare industry. Gamified services are more engaging and have the potential to drastically improve the preventative measures we put in place to head illnesses off at the pass.

I'm super excited to see whether gamification techniques can create lasting change – or whether it's simply a short term boost for motivation. Either way, it's pretty clear that gamification has its

---

[366] See: http://bit.ly/ageingstats

[367] See: http://bit.ly/aarpbabyboomers

place in the future of healthcare, and Dr. Bertalan Meskó has even identified four key use cases[368] in which gamification could be useful:

1. **Physical fitness**
2. **Medication and chronic condition management**
3. **Gamification for kids**
4. **Physical therapy and rehabilitation**

THE FUTURE OF CLINICAL TRIALS

You've probably already heard about Moore's law, the famous observation that Intel cofounder Gordon Moore made back in 1965. He noticed that the number of transistors per square inch on integrated circuit boards had doubled every year since their invention. As a general rule, this also means that processing power also doubles every year – or that devices can be halved in size.[369]

Eroom's law is a twist on that, and you might have noticed that 'Eroom' is 'Moore' spelled backwards. The idea here is that even with the continued advancements in technology, including the ever increasing processing power, storage and bandwidth capacities that are available to us, the R&D cost to bring a new drug to market is still increasing exponentially.

Fortunately, a solution could be forthcoming. Ryan Rossier, VP of platform solutions at healthcare consultancy Medullan, thinks that digitized clinical trials could be the answer. "The R&D departments within companies need to take advantage of new and evolving technological advancements," he explains. "We see this happening in other areas, like biology, where the cost to sequence your entire

---

[368] See: http://bit.ly/meskogamification

[369] See: http://bit.ly/mooreslawinvestopedia

genome will soon be less than a car payment."[370]

Rossier believes that digitized clinical trials could help to recruit and enroll patients faster while ensuring that they stick around for the full length of the trial. However, there are regulatory and compliance concerns and a general unwillingness from the industry to move towards new models. "For me," he says, "it's all about addressing the inertia in what is typically a very conservative industry. Although the experience level within companies is increasing, many drug developers remain cautious in their approach to digitizing clinical trials. Everyone seems to be taking a wait-and-see approach before investing heavily, seemingly not concerned with the downward trend of R&D efficiency. This creates a terrific opportunity to disrupt the status quo, show the value and benefit, and force others, including regulatory bodies, to be a part of the solution."

Other potential benefits of digitized clinical trials include:

REMOTE DATA COLLECTION: Removing the need for physical sites and reducing the need to collect and process data from case report forms.

REMOTE PATIENT MONITORING: Enabling patient monitoring outside of conventional clinical settings (e.g. in the home), which may increase access to care and compliance while simultaneously decreasing healthcare delivery costs.

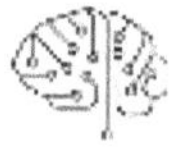

ONLINE EDUCATION: Learning materials can be provided to give trial participants a deeper understanding of their role in the research.

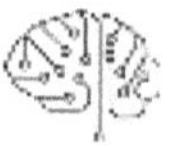

ADAPTIVE CLINICAL TRIALS: To capture, clean and interpret data in real-time to enable higher quality, shorter and less expensive clinical trials.

---

[370] See: http://bit.ly/digitizedclinicaltrials

Meanwhile, Stanford University welcomed a number of healthcare luminaries to campus as part of their Dean's Lecture Series, and the future of clinical trials was a hot topic. This was especially true during a talk by Thomas Pike, the former CEO of Quintiles.

"Healthcare reforms are needed," said Lloyd Minor MD, the dean of Stanford University School of Medicine, as he introduced Pike and his subject matter. "We need to rethink how we collect, secure, analyze and share our data. Fortunately, the opportunities surrounding us have never been better. Accomplishing this mission means embracing real-world evidence outside the clinical trial space."[371]

That's why Stanford itself is one of the leaders of Project Baseline, a four-year project in partnership with Google that aims to track the health of 10,000 participants over time. We talked some more about this in chapter twelve, when we took a look at the future of healthcare for insurers.

"This initiative will give us a view of health that is more of a feature film than it is a snapshot in time," Minor said. "[We'll see] not only what makes people sick, but also what keeps us healthy. It seeks to disrupt the status quo." Speaking as someone with prior experience in the field of clinical trials, I can wholeheartedly agree. It's just a matter of when digitized trials will become the new norm.

THE FUTURE OF HEALTHCARE

While I was carrying out my research for this book, I discovered an incredible tool from Focus Clinics that allows you to enter your age into an interactive graphic to see how old you'll be when major

---

[371] See: http://bit.ly/futureclinicaltrials

healthcare advances are likely to happen. I strongly suggest visiting the site and giving it a go so you can see when these technologies will arrive in your own lifetime.[372]

As of 2018, I'm 38-years-old, so let's use that as an example. By the time I'm 39, 3D printing will be a critical tool in over 35% of surgical procedures. By the time I'm 40, wearables will continuously monitor everything from posture to brain activity. In 2021, by the time I'm 41, AI will be routinely used to improve patient outcomes by providing clinical and medical solutions. By the time I'm 50, AI bots will outperform humans at some diagnoses and routine surgical procedures, while we'll be 3D printing synthetic tissues and entire human organs. And by the time I'm 55 – in just eighteen years – we'll be able to upgrade our senses with implants that detect more signals (such as radio waves and X-rays).

But while these futuristic devices and new technological applications will revolutionize the way we look at healthcare, the future is about outcomes and not just about the way we arrive at them. If we can use existing technologies in new ways then we'll do just that, as long as it benefits the patient.

Take text messaging (i.e. SMS). Despite being 25-years-old, text messaging is the fastest growing marketing channel in the UK and the rest of the world.[373] Marketers are notoriously obsessed with data and returns on investment, so the only reason why it's the fastest growing channel is that it works. If it's not broken, why fix it?

The same will happen in healthcare. Why shouldn't SMS be used? Why don't hospitals send SMS reminders to overweight patients reminding them to go to the gym? Why don't they text people to remind them it's time for a checkup or that they need to take their medication? Better still, SMS can be personalized, which leads us further towards a future in which personalized healthcare is the new norm.

Better still, no one ignores a text message – but they'll delete an

---

[372] See: http://bit.ly/futureofhealthcaretool

[373] See: http://bit.ly/smsfastestgrowing

email without even noticing it. That's because we get so much junk in our emails that we're used to just scanning them, whereas most of the SMS messages we receive are from friends and family – although I will admit that I ignore the messages from the juice bar that keeps texting me discount coupons. But if I flick through my phone and pick out a random text message, it's much more likely to be one that's important to me than if I do the same thing with my emails.

It's not all about tech. AI is taking away the mundane tasks that are driving costs and inefficiencies so that patients and doctors can have time to work together for better outcomes.

SMARTPHONE APPS

While I was writing *The Future of Healthcare,* I was forwarded some research from Carphone Warehouse which focused on the social impact of smartphones "including health, the economy, relationships, education and quality of life."[374] It was accompanied by a press release that claimed one in three Londoners have used an app to combat mental health problems. They also found that a quarter of Brits had used a smartphone app to improve their mental health, with the number rising to 44% of 25-34-year-olds.

One app user called Maddy (not her real name), who suffers from multiple personality disorder, anxiety, depression and a variety of other symptoms and conditions, said, "After the diagnosis, I was put on various medications to try to ease the symptoms. This required a lot of time off work as the physical side effects were hard to deal with. I was also on suicide watch for two months and was not allowed to be left alone under any circumstances. The doctors agreed that if my partner worked from home for the duration, I attended various regular intensive therapy sessions and used certain

---

[374] See: http://bit.ly/carphonewarehousestudy

smartphone apps, I could stay at home instead of being admitted to a psychiatric hospital and being under constant observation. Smartphones and technology have been criticized for being the cause of mental health problems and creating a world where everyone is addicted to technology. However, in my case, my smartphone kept me out of hospital."

When considering these applications, it helps to put them into perspective by examining the global mental health crisis. In 1990, 416 million people suffered from depression or anxiety worldwide, a figure which rose to 615 million in 2013. With around 10% of the world suffering from either depression or anxiety, it's easy to see how virtual support networks and cognitive behavioral therapy (CBT) apps could be deployed to improve patient outcomes. Now imagine if the data from these apps could be integrated with the rest of our medical records – and how useful it would be for busy physicians who don't get to see their patients as often as they might like to.

"These apps don't 'fix' mental health problems," Maddy says, "but they do help ease some of the problems associated with them. If it wasn't for the online community at Elsefriends and all of the other apps, I don't know if I would have been able to get through the darkest few months of my life."

WHY THE FUTURE MATTERS

One of the questions that I often hear when I talk about the future of healthcare is "why?" – why bother to usher in the future? And the same question applies to you as a reader. After all, we're all stakeholders in the future of healthcare, whether we're doctors and patients or whether we work for insurance or pharmaceutical companies. But why should we care about the future of healthcare? After all, I'm telling you right here and now that it's inevitable and that it *will* happen – so why should we go out of our way to help it along?

Perhaps the exact reason is different for each of us, but my personal feeling is that there's an enormous opportunity for us to improve lives. Anyone who's watched and/or experienced the healthcare industry for any amount of time is likely to share the frustration I feel at watching its inefficiencies.

Much of this was highlighted in a talk by Harvard Medical School Assistant Professor and CEO of Cyft, Dr. Leonard D'Avolio, at the HIMSS Big Data and Analytics Conference in Boston. "Every time you walk down the aisles of a supermarket, every item you're seeing, where it's positioned when it's restocked is dictated by the collective experience of shoppers who have come before you. Casinos know exactly when to offer their players some kind of incentive to keep them at the table a little longer. Even waste management has figured out how to optimize their routes to make sure that they're spending every dollar that they have as effectively as possible. They can predict landfill overfills years in advance."[375]

But the problem is that when it comes to healthcare, we're simply not using that information. D'Avolio points out that the most damning evidence of this comes from the number of people that die of preventable deaths each year. According to the Journal of Patient Safety, that number could be as high as 400,000 people per year who are dying due to accidents at the hands of the American healthcare system. That's the equivalent of a DC-10 and a Boeing 747 going down every day. D'Avolio says, "The third leading cause of death in America is preventable death at the hands of the healthcare system." And he believes that the very fact that we don't know the true figure is a damning indictment of the system. He says if something counts, we should count it.

I agree with D'Avolio, but I'd also add that even a single preventable death is one too many. After all, the clue is there in the name – it's preventable. He also argued that EMRs were initially developed specifically for the fee-for-service model and not for value-based care, which would explain the current fragmentation

---

[375] See: http://bit.ly/bigdatadavolio

and the fact that nobody has made major moves to address it. There needs to be an incentive for change to happen, and that incentive might have to be pressure from the general public. That means you, the reader, and it also means your friends and family. We all need to expect a better quality of healthcare. We all need to demand the future and raise our voices until legislators – and innovators – take action.

If nothing else, we have a responsibility to the next generation to make this happen.

HUMANS AND MACHINES PARTNERING FOR BETTER OUTCOMES

The future, then, is all about humans and machines partnering for better outcomes. It'll be cheaper, more effective, and will dramatically improve our overall quality of life.

It's easy to see how healthcare tech will make its way into the field of professional sports. Just look at 94Fifty, a smart basketball that "looks and performs like a standard basketball while unobtrusively gathering information about shooting arc, dribble intensity and speed, shot release speed and shot backspin during both indoor and outdoor usage." It even allows people to compete with each other via Twitter. And, as Anita Berryman explains for PTC, "[It] provides actionable feedback to the player to help improve their fundamental basketball skills."[376]

The main observation here is that technologies are starting to blur the lines between healthcare and entertainment. Devices like the Fitbit and the Apple Watch are consumer electronics devices as much as they are healthcare devices. In the future, apps like Babylon will use new technologies to replace the first line of diagnosis for

---

[376] See: http://bit.ly/smartbasketballiot

when people have minor illnesses and injuries like cuts or colds. Doctors will still need to oversee things like prescriptions, but new apps will help to free up staff and cut down on costs, making the healthcare industry as a whole exponentially more profitable.

Imagine an app that showed users all of the doctors within their insurance network and which enabled them to narrow them down by their specialisms. Imagine if that same app allowed them to arrange a telehealth consultation and if it could track and share the patient's healthcare information from all of their providers and healthcare devices. Imagine if it included 60-second documentaries, disease education and discussion forums, not to mention support groups for people with similar illnesses. Imagine if all of the data it gathered could be aggregated and analyzed anonymously to identify regional, national and global healthcare trends. Imagine if that same app could be used to educate people about clinical trials and to recruit new participants.

The technology to do this is already here. The future of healthcare is coming. It's only a matter of time.

ACKNOWLEDGEMENTS

WHERE DO I BEGIN?

First off, thanks to my beloved grandmother Jacqueline Nkamanyi (Mami Jaco) for teaching me the importance of education. R.I.P. Mami Jaco. You were my biggest inspiration and your belief in me lives on in everything I do.

Special thanks to my family (Martina, Yves, Elise, Irene, Anne Marie and Mario) for all the support, patience and guidance while writing this book. Thanks also to my mother (Dr. Regina Nkamanyi) for shaping my medical career and to Anil Moolchandani from Cornell University for first introducing me to the power of big data and artificial intelligence.

I'd also like to thank everyone who took the time to talk to me throughout the creation of this book. Thanks also to my publishing team, starting with Dane Cobain, my editor, who helped me to shape my manuscript into the book that you're holding in your hands. Thanks are also due to Martina Milova for handling photography and Alex Bashta for the illustrations. Sean Strong did a fantastic job with the cover art, too.

Finally, thanks to you, the reader. The future of healthcare is coming – but only if you help to make it happen.

Over to you.

ABOUT THE AUTHOR

EMMANUEL FOMBU, MD, MBA, is an Ivy League educated physician, author, speaker and healthcare executive turned Silicon Valley entrepreneur. He is a medical futurist, an advocate for value-based healthcare and the 2017 winner of the prestigious New York City Health Business Leaders Boldest Digital Health Influencer Award.

As a medical futurist, Dr. Fombu champions the potential for the internet of things, AI and machine learning to revolutionize the healthcare industry. He's passionate about m-health, personalized medicine, genomics, nanotechnology, big data, artificial intelligence, machine learning, the internet of things and digital medicine. He serves as an external advisory board member on the Massachusetts Institute of Technology's MIT.nano project.

Dr. Fombu trained at Emory-Crawford Long Hospital and holds an MBA from Cornell University's Johnson School of Business and a certification on artificial intelligence from MIT's Computer Science and Artificial Intelligence Lab. He's taught allied health students, founded a health and education focused non-profit organization and advocates for patients with rare and chronic diseases. Dr Fombu works extensively with venture capitalists and start-ups in digital health and biotech.

In addition to *The Future of Healthcare: Humans and Machines Partnering for Better Outcomes,* Dr. Fombu has published multiple scientific papers in world renowned, peer reviewed scientific journals. He lives in New York City.

JOIN THE CONVERSATION

Thanks for reading *The Future of Healthcare: Humans and Machines Partnering for Better Outcomes*! I hope it's inspired you to take your healthcare into your own hands and to help usher in a new era in global medicine.

Whether you loved the book or you hated it, I want to know what you think. Every review helps to make the future of healthcare a reality, so please do take the time to leave a short review on Amazon and Goodreads and to spread the word amongst your friends and family.

Want to get involved in the discussion? Follow me on LinkedIn, Facebook and Twitter or join the conversation using the #FutureOfHealthcare hashtag.

For more information:

emmanuelfombu.com
twitter.com/fombumd
facebook.com/fombumd
bit.ly/fombulinkedin

CPSIA information can be obtained
at www.ICGtesting.com
Printed in the USA
LVHW112135110319
610305LV00012B/472/P